ADAPTIVE ARRAY ANTENNA FOR

MOBILE & SATELLITE COMMUNICATION

MAINAK MUKHOPADHYAY

Contents

CHAPTER 4

Direct Data Domain Least Squares approaches to Adaptive Processing based on Single Snapshots of Data.

CHAPTER 5

Conjugate Gradient Method for D³LS Algorithm

CHAPTER 6

Estimation of the Direction of Arrival Using the Matrix Pencil Method

References

[1] "Special issue on active and adaptive antennas," IEEE Trans. Antennas Propagat., vol. AP-12, Mar. 1964.

[2] "Special issue on adaptive antennas," IEEE Trans. Antennas Propagat., vol. AP-24, Sept. 1976.

[3] "Special issue on adaptive processing antenna systems," IEEE Trans. Antennas Propagat., vol. AP-34, Mar. 1986.

[4] "Special issue on adaptive systems and applications," IEEE Trans. Circuits Syst., vol. CAS-34, July 1987.

[5] "Special issue on beamforming," IEEE J. Oceanic Eng., vol. OE-10, July 1985.

[6] "Special issue on underwater acoustic signal processing," IEEE J. Oceanic Eng., vol. OE-12, Jan. 1987.

[7] J. E. Hudson, Adaptive Array Principles. London: Peregrinus, 1981.

[8] R. A. Monzingo and T. W. Miller, Introduction to Adaptive Arrays. New York: Wiley, 1980.

[9] S. Haykin, Ed., Array Signal Processing. Englewood Cliffs, NJ: Prentice-Hall, 1985.

[10] B. Widrow and S. D. Stearns, Adaptive Signal Processing. Englewood Cliffs, NJ: Prentice-Hall, 1985.

[11] R. T. Compton, Jr., Adaptive Antennas: Concepts and Performances, Englewood Cliffs, NJ: Prentice-Hall, 1988.

[12] M. T. Ma, Theory and Application of Antenna Arrays. New York: Wiley, 1974.

[13] J. D. Marr, "A selected bibliography on adaptive antenna arrays," IEEE Trans. aerosp. Electron. Syst., vol. AES-22, pp. 781–788, 1986.

[14] K. Takao, M. Fujita, and T. Nishi, "An adaptive antenna array under directional constraint," IEEE Trans. Antennas Propagat., vol. AP-24, pp. 662–669, 1976.

[15] L. Stark, "Microwave theory of phased-array antennas—A review," Proc. IEEE, vol. 62, pp. 1661–1701, 1974.

[16] N. L. Owsley, "A recent trend in adaptive spatial processing for sensor arrays: Constrained adaptation," in Signal Processing, J. W. R. Griffiths et al., Eds. New York: Academic, 1973.

[17] A. M. Vural, "An overview of adaptive array processing for sonar application," EASCON'75 Rec., pp. 34A–34M.

[18] P. M. Schultheiss, "Some lessons from array processing theory," in Aspects of Signal Processing, part 1, G. Tacconi, Ed. Dordrecht-Holland: Reidel, 1977, pp. 309–331.

[19] H. A. d'Assumpcao, "Some new signal processors for array of sensors," IEEE Trans. Inform. Theory, vol. IT-26, pp. 441–453, 1980.

[20] R. J. Maillous, "Phased array theory and technology," Proc. IEEE, vol. 70, pp. 246–291, 1982.

[21] S. Haykin, J. P. Reilly, V. Kezys, and E. Vertatschitsch, "Some aspects of array signal processing," Proc. Inst. Elect. Eng., vol. 139, pt. F, pp. 1–19, 1992.

[22] H. A. d'Assumpcao and G. E. Mountford, "An overview of signal processing for arrays of receivers," J. Inst. Eng. Aust. and IREE Aust., vol. 4, pp. 6–19, 1984.

[23] B. D. Van Veen and K. M. Buckley, "Beamforming: A versatile approach to spatial filtering," IEEE Aerosp. Electron. Syst. Mag., vol. 5, pp. 4–24, 1988.

[24] S. P. Applebaum, "Adaptive arrays," IEEE Trans. Antennas Propagat., vol. AP-24, pp. 585–598, 1976.

[25] O. L. Frost III, "An algorithm for linearly constrained adaptive array processing," Proc. IEEE, vol. 60, pp. 926–935, 1972.

[26] W. F. Gabriel, "Adaptive arrays—An introduction," Proc. IEEE, vol. 64, pp. 239–272, 1976.

[27] B. Widrow, P. E. Mantey, L. J. Griffiths, and B. B. Goode, "Adaptive antenna systems," Proc. IEEE, vol. 55, pp. 2143–2158, 1967.

[28] B. Widrow, J. R. Glover, J. M. McCool, J. Kaunitz, C. S. Williams, R. H. Hearn, J. R. Zeidler, E. Dong, Jr., and R. C. Goodlin, "Adaptive noise canceling: Principles and applications," Proc. IEEE, vol. 63, pp. 1692–1716, 1975.

[29] L. C. Godara, "Error analysis of the optimal antenna array processors," IEEE Trans. Aerosp. Electron. Syst., vol. AES-22, pp. 395–409, 1986.

[30] H. Krim and M. Viberg, "Two decades of array signal processing: The parametric approach," IEEE Signal Processing Mag., pp. 67–94, July 1996.

[31] J. Munier and G. Y. Delisle, "Spatial analysis in passive listening using adaptive techniques," Proc. IEEE, vol. 75, pp. 1458–1471, 1987.

[32] S. Sivanand, "On adaptive arrays in mobile communication," in Proc. IEEE National Telesystems Conf., Atlanta, GA, 1993, pp. 55–58.

[33] V. C. Anderson and P. Rudnick, "Rejection of a coherent arrival at an array," J. Acoust. Soc. Amer., vol. 45, pp. 406–410, 1969.

[34] V. C. Anderson, "DICANNE, a realizable adaptive process," J. Acoust. Soc. Amer., vol. 45, pp. 398–405, 1969.

[35] L. C. Godara and A. Cantoni, "Uniqueness and linear independence of steering vectors in array space," J. Acoust. Soc. Amer., vol. 70, pp. 467–475, 1981.

[36] Y. Bresler, V. U. Reddy, and T. Kailath, "Optimum beamforming for coherent signal and interferences," IEEE Trans. Acoust., Speech, Signal Processing, vol. 36, pp. 833–843, 1988.

[37] S. Choi, T. K. Sarkar, and S. S. Lee, "Design of twodimensional Tseng window and its application to antenna array for the detection of AM signal in the presence of strong jammers in mobile communications," Signal Process., vol. 34, pp. 297–310, 1993.

[38] I. Chiba, T. Takahashi, and Y. Karasawa, "Transmitting null beam forming with beam space adaptive array antennas," in Proc. IEEE 44th Vehicular Technology Conf., Stockholm, Sweden, 1994, pp. 1498–1502.

[39] B. Friedlander and B. Porat, "Performance analysis of a nullsteering algorithm based on direction-of-arrival estimation," IEEE Trans. Acoust., Speech, Signal Processing, vol. 37, pp. 461–466, 1989.

[40] I. S. Reed, J. D. Mallett, and L. E. Brennan, "Rapid convergence rate in adaptive arrays," IEEE Trans. Aerosp. Electron. Syst., vol. AES-10, pp. 853–863, 1974.

[41] L. E. Brennan and I. S. Reed, "Theory of adaptive radar," IEEE Trans. Aerosp. Electron. Syst., vol. AES-9, pp. 237–252, 1973.

[42] H. Cox, "Resolving power and sensitivity to mismatch of optimum array processors," J. Acoust. Soc. Amer., vol. 54, pp. 771–785, 1973.

[43] J. Capon, "High-resolution frequency-wave number spectrum analysis," Proc. IEEE, vol. 57, pp. 1408–1418, 1969.

[44] I. J. Gupta and A. A. Ksienski, "Dependence of adaptive array performance on conventional array design," IEEE Trans. Antennas Propagat., vol. AP-30, pp. 549–553, 1982.

[45] I. J. Gupta, "Effect of jammer power on the performance of adaptive arrays," IEEE Trans. Antennas Propagat., vol. AP-32, pp. 933–934, 1984.

[46] J. H. Winters, "Optimum combining in digital mobile radio with cochannel interference," IEEE J. Select. Areas Commun., vol. SAC-2, pp. 528–539, 1984.

[47] J. H. Winters, "Optimum combining for indoor radio systems with multiple users," IEEE Trans. Commun., vol. COM-35, pp. 1222–1230, 1987.

[48] B. Suard, A. F. Naguib, G. Xu, and A. Paulraj, "Performance of CDMA mobile communication systems using antenna arrays," IEEE Int. Conf. Acoustics, Speech, and Signal Processing (ICASSP), Minneapolis, MN, 1993, pp. 153–156.

[49] A. F. Naguib and A. Paulraj, "Performance of CDMA cellular networks with base-station antenna arrays," in Proc. IEEE Int. Zurich Seminar on Communications, 1994, pp. 87–100.

[50] C. L. Zahm, "Application of adaptive arrays to suppress strong jammers in the presence of weak signals," IEEE Trans. Aerosp. Electron. Syst., vol. AES-9, pp. 260–271, 1973.

[51] S. P. Applebaum and D. J. Chapman, "Adaptive arrays with main beam constraints," IEEE Trans. Antennas Propagat., vol. AP-24, pp. 650–662, 1976.

[52] B. Widrow, K. M. Duvall, R. P. Gooch, and W. C. Newman, "Signal cancellation phenomena in adaptive antennas: Causes and cures," IEEE Trans. Antennas Propagat., vol. AP-30, pp. 469–478, 1982.

[53] R. T. Compton Jr., "The power-inversion adaptive array: Concepts and performance," IEEE Trans. Aerosp. Electron. Syst., vol. AES-15, pp. 803–814, 1979.

[54] L. J. Griffiths, "A comparison of multidimensional Weiner and maximum-likelihood filters for antenna arrays," in Proc. IEEE, vol. 55, pp. 2045–2047, 1967.

[55] A. Flieller, P. Larzabal, and H. Clergeot, "Applications of high resolution array processing techniques for mobile communication system," in Proc. IEEE Intelligent Vehicles Symp., Paris, France, 1994, pp. 606–611.

[56] J. H. Winters, J. Salz, and R. D Gitlin, "The impact of antenna diversity on the capacity of wireless communication systems," IEEE Trans. Commun., vol. 42, pp. 1740–1751, 1994.

[57] S. Choi and T. K. Sarkar, "Adaptive antenna array utilizing the conjugate gradient method for multipath mobile communication," Signal Process., vol. 29, pp. 319–333, 1992.

[58] S. Anderson, M. Millnert, M. Viberg, and B. Wahlberg, "An adaptive array for mobile communication systems," IEEE Trans. Veh. Technol., vol. 40, pp. 230–236, 1991.

[59] T. Gebauer and H. G. Gockler, "Channel-individual adaptive beamforming for mobile satellite communications," IEEE J. Select. Areas Commun., vol. 13, pp. 439–448, 1995.

[60] J. F. Diouris, B. Feuvrie, and J. Saillard, "Adaptive multisensor receiver for mobile communications," Ann. Telecommun., vol. 48, pp. 35–46, 1993.

[61] P. W. Howells, "Explorations in fixed and adaptive resolution at GE and SURC," IEEE Trans. Antennas Propagat., vol. AP-24, pp. 575–584, 1976.

[62] L. J. Griffiths and C. W. Jim, "An alternative approach to linearly constrained adaptive beamforming," IEEE Trans. Antennas Propagat., vol. AP-30, pp. 27–34, 1982.

[63] L. J. Griffiths, "An adaptive beamformer which implements constraints using an auxiliary array processor," in Aspects of Signal Processing, part 2, G. Tacconi, Ed. Dordrecht-Holland: Reidel, 1977, pp. 517–522.

[64] C. W. Jim, "A comparison of two LMS constrained optimal array structures," Proc. IEEE, vol. 65, pp. 1730–1731, 1977.

[65] A. Cantoni and L. C. Godara, "Fast algorithms for time domain broadband adaptive array processing," IEEE Trans. Aerosp. Electron. Syst., vol. AES-18, pp. 682–699, 1982.

[66] B. D. Van Veen and R. A. Roberts, "Partially adaptive beamformer design via output power minimization," IEEE Trans. Acoust., Speech, Signal Processing, vol. ASSP-35, pp. 1524–1532, 1987.

[67] B. D. Van Veen, "An analysis of several partially adaptive beamformer designs," IEEE Trans. Acoust., Speech, Signal Processing, vol. 37, pp. 192–203, 1989.

[68] , "Optimization of quiescent response in partially adaptive beamformers," IEEE Trans. Acoust., Speech, Signal Processing, vol. 38, pp. 471–477, 1990.

[69] F. Qian and B. D. Van Veen, "Partially adaptive beamformer design subject to worst case performance constraints," IEEE Trans. Signal Processing, vol. 42, pp. 1218–1221, 1994.

[70] , "Partially adaptive beamforming for correlated interference rejection," IEEE Trans. Signal Processing, vol. 43, pp. 506–515, 1995.

[71] D. J. Chapman, "Partially adaptivity for the large array," IEEE Trans. Antennas Propagat., vol. AP-24, pp. 685–696, 1976.

[72] D. R. Morgan, "Partially adaptive array techniques," IEEE Trans. Antennas Propagat., vol. AP-26, pp. 823–833, 1978.

[73] A. Cantoni and L. C. Godara, "Performance of a postbeamformer interference canceller in the presence of broadband directional signals," J. Acoust. Soc. Amer., vol. 76, pp. 128–138, 1984.

[74] L. C. Godara and A. Cantoni, "The effect of bandwidth on the performance of post beamformer interference canceller," J. Acoust. Soc. Amer., vol. 80, pp. 794–803, 1986.

[75] L. C. Godara, "Analysis of transient and steady state weight covariance in adaptive postbeamformer interference canceller,"J. Acoust. Soc. Amer., vol. 85, pp. 194–201, 1989.

[76] , "Postbeamformer interference canceller with improved performance," J. Acoust. Soc. Amer., vol. 85, pp. 202–213, 1989.

[77] , "Adaptive postbeamformer interference canceller with improved performance in the presence of broadband directional sources," J. Acoust. Soc. Amer., vol. 89, pp. 266–273, 1991.

[78] E. Brookner and J. M. Howell, "Adaptive-adaptive array processing," Proc. IEEE, 1986, vol. 74, pp. 602–604.

[79] J. T. Mayhan, "Adaptive nulling with multiple beam antennas," IEEE Trans. Antennas Propagat., vol. AP-26, pp. 267–273, 1978.

[80] R. Klemm, "Suppression of jammers by multiple beam signal processing," in Proc. IEEE Int. Radar Conf., Sendai, Japan, 1975, pp. 176–180.

[81] J. Gobert, "Adaptive beam weighting," IEEE Trans. Antennas Propagat., vol. AP-24, pp. 744–749, 1976.

[82] C. L. Dolf, "A current distribution for broadside arrays which optimizes the relationship between beamwidth and sidelobe levels," Proc. IRE, vol. 34, pp. 335–348, 1946.

[83] L. J. Griffiths and K. M. Buckley, "Quiescent pattern control in linearly constrained adaptive arrays," IEEE Trans. Acoust., Speech, Signal Processing, vol. ASSP-35, pp. 917–926, 1987.

[84] C. Y. Tseng and L. J. Griffiths, "A simple algorithm to achieve desired patterns for arbitrary arrays," IEEE Trans. Signal Processing, vol. 40, pp. 2737–2746, 1992.

IV

[85] R. J. Webster and T. N. Lang, "Prescribed sidelobes for the constant beam width array," IEEE Trans. Acoust., Speech, Signal Processing, vol. 38, pp. 727–730, 1990.

[86] M. H. Er, S. L. Sim, and S. N. Koh, "Application of constrained optimization techniques to array pattern synthesis," Signal Process., vol. 34, pp. 327–334, 1993.

[87] M. Simaan, "Optimum array filters for array data signal processing," IEEE Trans. Acoust., Speech, Signal Processing, vol. ASSP-31, pp. 1006–10015, 1983.

[88] D. E. N. Davies, "Independent angular steering of each zero of the directional pattern for a linear array," IEEE Trans. Antennas Propagat., vol. AP-15, pp. 296–298, 1967.

[89] B. D. Van Veen, "Adaptive convergence of linearly constrained beamformers based on the sample covariance matrix," IEEE Trans. Signal Processing, vol. 39, pp. 1470–1473, 1991.

[90] L. C. Godara, "A robust adaptive array processor," IEEE Trans. Circuits Syst., vol. CAS-34, pp. 721–730, 1987.

[91] N. K. Jablon, "Adaptive beamforming with the generalized sidelobe canceller in the presence of array imperfections," IEEE Trans. Antennas Propagat., vol. AP-34, pp. 996–1012, 1986.

[92] W. F. Gabriel, "Using spectral estimation techniques in adaptive processing antenna systems," IEEE Trans. Antennas Propagat., vol. AP-34, pp. 291–300, 1986.

[93] Y. L. Su, T. J. Shan, and B. Widrow, "Parallel spatial processing: A cure for signal cancellation in adaptive arrays," IEEE Trans. Antennas Propagat., vol. AP-34, pp. 347–355, 1986.

[94] P. Kawala and U. H. Sheikh, "Adaptive multiple-beam array for wireless communications," in Proc. Inst. Elect. Eng. 8th Int. Conf. Antennas and Propagation, Edinburgh, Scotland, 1993, pp. 970–974.

[95] W. Chujo and K. Yasukawa, "Design study of digital beam forming antenna applicable to mobile satellite communications," IEEE Antennas and Propagation Symp. Dig., Dallas, TX, pp. 400–403, 1990.

[96] I. Chiba, W. Chujo, and M. Fujise, "Beamspace constant modulus algorithm adaptive array antennas," in Proc. Inst. Elect. Eng. 8th Int. Conf. Antennas and Propagation, Edinburgh, Scotland, 1993, pp. 975–978.

[97] M. A. Jones and M. A. Wickert, "Direct sequence spread spectrum using directionally constrained adaptive beamforming to null interference," IEEE J. Select. Areas Commun., vol. 13, pp. 71–79, 1995.

[98] S. Sakagami, S. Aoyama, K. Kuboi, S. Shirota, and A. Akeyama, "Vehicle position estimates by multibeam antennas in multipath environments," IEEE Trans. Veh. Technol., vol. 41, pp. 63–68, 1992.

[99] T. Tanaka, R. Miura, I. Chiba, and Y. Karasawa, "An ASIC implementation scheme to realize a beam space CMA adaptive array antenna," IEICE Trans. Commun., vol. E78-B, pp. 1467–1473, Nov. 1995.

[100] W. E. Rodgers and R. T. Compton Jr., "Adaptive array bandwidth with tapped delay line processing," IEEE Trans. Aerosp. Electron. Syst., vol. AES-15, pp. 21–28, 1979.

[101] J. T. Mayhan, A. J. Simmons, and W. C. Cummings, "Wideband adaptive nulling using tapped delay lines," IEEE Trans. Antennas Propagat., vol. AP-29, pp. 923–936, 1981.

[102] E. W. Vook and R. T. Compton, Jr., "Bandwidth performance of linear arrays with tapped delay line processing," IEEE Trans. Aerosp. Electron. Syst., vol. 28, pp. 901–908, 1992.

[103] R. T. Compton, Jr., "The bandwidth performance of a two element adaptive array with tapped delay line processing," IEEE Trans. Antennas Propagat., vol. 36, pp. 5–14, 1988.

[104] C. C. Ko, "Jamming rejection capability of broadband Frost power inversion array," Proc. Inst. Elect. Eng., vol. 128, pp. pt. F, 140–151, 1981.

[105] , "Tracking performance of a broadband tapped delay line adaptive array using the LMS algorithm," Proc. Inst. Elect. Eng., vol. 134, pt. F, pp. 295–302, 1987.

[106] D. Nunn, "Performance assessments of a time domain adaptive processor in a broadband environment," Proc. Inst. Elect. Eng., Pts. F and H, vol. 130, pp. 139–145, 1983.

[107] C. C. Yeh, Y. J. Hong, and D. R. Ucci, "Use of tapped delay line adaptive array to increase the number of degrees of freedom for interference suppression," IEEE Trans. Aerosp. Electron. Syst., vol. AES-23, pp. 809–813, 1987.

[108] K. K. Scott, "Transversal filter techniques for adaptive array applications," Proc. Inst. Elect. Eng., Pts. F and H, vol. 130, pp. 29–35, 1983.

[109] T. S. Durrani, N. L. M. Murukutla, and K. C. Sharman, "Constrained algorithm for multi-input adaptive latices in array processing," in Proc. ICASSP, Atlanta, GA, 1981, pp. 297–301.

[110] D. Alexandrou, "Boundary reverberation rejection via constrained adaptive beamforming," J. Acoust. Soc. Amer., vol. 82, pp. 1274–1290, 1987.

[111] F. Ling, D. Manolakis, and J. G. Proakis, "Numerically robust least-squares lattice ladder algorithms with direct updating of the reflection coefficients," IEEE Trans. Acoust., Speed, Signal Processing, vol. ASSP-34, pp. 837–845, 1986.

[112] Y. Iiguni, H. Sakai, and H. Tokumaru, "Convergence properties of simplified gradient adaptive lattice algorithms," IEEE Trans. Acoust., Speech, Signal Processing, vol. ASSP-33, pp. 1427–1434, 1985.

[113] G. R. L. Sohie and L. H. Sibul, "Stochastic convergence properties of the adaptive gradient lattice," IEEE Trans. Acoust., Speech, Signal Processing, vol. ASSP-32, pp. 102–107, 1984.

[114] M. H. Er and A. Cantoni, "Derivative constraints for broadband element space antenna array processors," IEEE Trans. Acoust., Speech, Signal Processing, vol. ASSP-31, pp. 1378–1393, 1983.

[115] M. H. Er and B. P. Ng, "On derivative constrained broadband beamforming," IEEE Trans. Acoust., Speech, Signal Processing, vol. 38, pp. 551–552, 1990.

[116] I. Thng, A. Cantoni, and Y. H. Leung, "Derivative constrained optimum broadband antenna arrays," IEEE Trans. Signal Processing, vol. 41, pp. 2376–2388, 1993.

[117] K. Takao and T. Ishizaki, "Constraints of the output power minimization adaptive array for broadband desired signal," Trans. Inst. Electr. Commun. Eng. Jpn. B, vol. J68-B, pp. 411–418, 1985.

[118] K. M. Buckley, "Spatial/spectral filtering with linearly constrained minimum variance beamformers," IEEE Trans. Acoust., Speech, Signal Processing, vol. ASSP-35, pp. 249–266, 1987.

[119] K. M. Ahmed and R. J. Evans, "Broadband adaptive array processing," Proc. Inst. Elect. Eng., vol. 130, pt. F, pp. 433–440, 1983.

[120] M. H. Er, "On the limiting solution of quadratically constrained broadband beamformers," IEEE Trans. Signal Processing, vol. 41, pp. 418–419, 1993.

[121] K. M. Ahmed and R. J. Evans, "An adaptive array processor with robustness and broadband capabilities," IEEE Trans. Antennas Propagat., vol. AP-32, pp. 944–950, 1984.

[122] N. Kikuma and K. Takao, "Broadband and robust adaptive antenna under correlation constraints," Proc. Inst. Elect. Eng., vol. 136, pt. H, pp. 85–89, 1989.

[123] C. L. B. Despins, D. D. Falconer, and S. A. Mahmoud, "Compound strategies of coding, equalization and space diversity for wideband TDMA indoor wireless channels," IEEE Trans. Veh. Technol., vol. 41, pp. 369–379, 1992.

[124] N. Ishii and R. Kohno, "Spatial and temporal equalization based on an adaptive tapped-delay-line array antenna," IEICE Trans. Commun., vol. E78-B, pp. 1162–1169, Aug. 1995.

[125] R. Kohno, H. Wang, and H. Imai, "Adaptive array antenna combined with tapped delay line using processing gain for spread spectrum CDMA systems," presented at the IEEE Int. Symp. Personal Indoor and Mobile Radio Communications, Boston, MA, 1992.

[126] M. H. Er and A. Cantoni, "An unconstrained partitioned realization for derivative constrained broadband antenna array processors," IEEE Trans. Acoust., Speech, Signal Processing, vol. ASSP-34, pp. 1376–1379, 1986.

[127] L. P. Winkler and M. Schwartz, "Adaptive nonlinear optimization of the signal-to-noise ratio of an array subject to a constraint," J. Acoust. Soc. Amer., vol. 52, pp. 39–51, 1972.

[128] C. C. Ko, "Fast null steering algorithm for broadband power inversion array," Proc. Inst. Elect. Eng., vol. 137, pt. F, pp. 377–383, 1990.

[129] K. Takao and K. Komiyama, "An adaptive antenna for rejection of wideband interference," IEEE Trans. Aerosp. Electron. Syst., vol. AES-16, pp. 452–459, 1980.

[130] K. C. Huang, S. H. Chang, and Y. H. Chen, "An alternative structure for adaptive broadband beamforming with imperfect arrays," J. Acoust. Soc. Amer., vol. 87, pp. 1218–1226, 1990.

[131] S. J. Chern and C. Y. Sung, "The hybrid Frost's beamforming algorithm for multiple jammers suppression," Signal Process., vol. 43, pp. 113–132, 1995.

[132] R. W. Harris, D. M. Chabriew, and F. A. Bishop, "A variable step size (VS) adaptive filter algorithm," IEEE Trans. Acoust., Speech, Signal Processing, vol. ASSP-34, pp. 309–316, 1986.

[133] R. Y. Chen and C. L. Wang, "On the optimum step size for the adaptive sign and LMS algorithms," IEEE Trans. Circuits Syst., vol. 37, pp. 836–840, 1990.

[134] S. Nordebo, I. Claesson, and S. Nordhom, "Weighted Tschebysheff approximation for the design of broadband beamformers using quadratic programming," IEEE Signal
Processing Lett., vol. 1, pp. 103–105, 1994.

[135] S. Valaee and P. Kabal, "Wideband array processing using a two-sided correlation transformation," IEEE Trans. Signal Processing, vol. 43, pp. 160–172, 1995.

[136] W. S. Hodgkiss, "Adaptive array processing: Time vs. frequency domain," ICASSP: Int. Conf. Acoustic, Speech, Signal Processing, Washington, D.C., 1979, pp. 282–284.

[137] L. Armijo, W. Daniel, and W. M. Labuda, "Applications of the FFT to antenna array beamforming," in EASCON, Rec. IEEE Electronics and Aerospace Systems Conv., Washington, D.C., 1974, pp. 381–383.

[138] M. Dentino, J. McCool, and B. Widrow, "Adaptive filtering in frequency domain," Proc. IEEE, vol. 66, pp. 1658–1659, 1978.

[139] S. S. Narayan and A. M. Peterson, "Frequency domain leastmean- square algorithm," Proc. IEEE, vol. 69, pp. 124–126, 1981.

[140] M. E. Weber and R. Heisler, "A frequency-domain beamforming algorithm for wideband, coherent signal processing," J. Acoust. Soc. Amer., vol. 76, pp. 1132–1144, 1984.

[141] J. J. Shynk and R. P. Gooch, "Frequency-domain adaptive pole-zero filtering," Proc. IEEE, vol. 73, pp. 1526–1528, 1985.

[142] S. Florian and N. J. Bershad, "A weighted normalized frequency domain LMS adaptive algorithm," IEEE Trans. Acoust., Speech, Signal Processing, vol. 36, pp. 1002–1007, 1988.

[143] N. J. Bershad and P. L. Feintuch, "A normalized frequency domain LMS adaptive algorithm," IEEE Trans. Acoust., Speech, Signal Processing, vol. ASSP-34, pp. 452–461, 1986.

[144] F. A. Reed, P. L. Feintuch, and N. J. Bershad, "The application of the frequency domain LMS adaptive filter to split array bearing estimation with a sinusoidal signal," IEEE Trans. Acoust., Speech, Signal Processing, vol. ASSP-33, pp. 61–69, 1985.

[145] D. Mansour and A. H. Gray Jr., "Unconstrained frequency domain adaptive filter," IEEE Trans. Acoust., Speech, Signal Processing, vol. ASSP-30, pp. 168–170, 1982.

[146] R. Kumaresan, "On a frequency domain analog of Prony's method," IEEE Trans. Acoust., Speech, Signal Processing, vol. 38, pp. 168–170, 1990.

[147] G. A. Clark, S. R. Parker, and S. K. Mitra, "A unified approach to time-and frequency-domain realization of FIR adaptive digital filters," IEEE Trans. Acoust., Speech, Signal Processing, vol. ASSP-31, pp. 1073–1083, 1983.

[148] J. X. Zhu and H. Wang, "Adaptive beamforming for correlated signal and interference: A frequency domain smoothing approach," IEEE Trans. Acoust., Speech, Signal Processing, vol. 38, pp. 193–195, 1990.

[149] L. C. Godara, "Application of the fast Fourier transform to broadband beamforming," J. Acoust. Soc. Amer., vol. 98, pp. 230–240, 1995.

[150] M. J. Hinich, "Frequency-wave number array processing," J. Acoust. Soc. Amer., vol. 69, pp. 732–737, 1981.

[151] V. C. Anderson, "Digital array phasing," J. Acoust. Soc. Amer., vol. 32, pp. 867–870, 1960.

[152] R. G. Pridham and R. A. Mucci, "A novel approach to digital beamforming," J. Acoust. Soc. Amer., vol. 63, pp. 425–434, 1978.

[153] P. Barton, "Digital beamforming for radar," Proc. Inst. Elect. Eng., vol. 127, pt. F, pp. 266–277, 1980.

[154] N. J. Mohamed, "Two-dimensional beamforming with nonsinusoidal signals," IEEE Trans. Electromag. Compat., vol. EMC-29, pp. 303–313, 1987.

[155] D. E. Dudgeon, "Fundamentals of digital array processing," Proc. IEEE, vol. 65, pp. 898–904, 1977.

[156] R. A. Mucci, "A comparison of efficient beamforming algorithms," IEEE Trans. Acoust., Speech, Signal Processing, vol. ASSP-32, pp. 548–558, 1984.

[157] R. G. Pridham and R. A. Mucci, "Digital interpolation beamforming for low-pass and bandpass signals," Proc. IEEE, vol. 67, pp. 904–919, 1979.

[158] H. Fan, E. I. El-Masry, and W. K. Jenkins, "Resolution enhancement of digital beamforming," IEEE Trans. Acoust., Speech, Signal Processing, vol. ASSP-32, pp. 1041–1052, 1984.

[159] B. Maranda, "Efficient digital beamforming in the frequency domain," J. Acoust. Soc. Amer., vol. 86, pp. 1813–1819, 1989.

[160] P. Rudnick, "Digital beamforming in the frequency domain," J. Acoust. Soc. Amer., vol. 46, pp. 1089–1090, 1969.

[161] R. A. Gabel and R. R. Kurth, "Hybrid time-delay/phase-shift digital beamforming for uniform collinear array," J. Acoust. Soc. Amer., vol. 75, pp. 1837–1847, 1984.

[162] J. J. Brady, "A serial phase shift beamformer using charge transfer devices," J. Acoust. Soc. Amer., vol. 68, pp. 504–506, 1980.

[163] P. D. Sylva, P. Menard, and D. Roy, "A reconfigurable realtime interpolation beamformer," IEEE J. Oceanic Eng., vol. OE-11, pp. 123–126, 1986.

[164] G. J. DeMuth, "Frequency domain beamforming techniques," in Proc. IEEE Int. Conf. Acoustics, Speech, Signal Processing (ICASSP), 1977, pp. 713–715.

[165] A. Papoulis, "A new algorithm in spectral analysis and bandlimited extrapolation," IEEE Trans. Circuits Syst., vol. CAS-22, pp. 735–742, 1975.

[166] B. J. Sullivan, "Effect of sampling rate on the conjugate gradient method applied to signal extrapolation," IEEE Trans. Signal Processing, vol. 39, pp. 1235–1238, 1991.

[167] J. A. Cadzow, "An extrapolation procedure for band-limited signals," IEEE Trans. Acoust., Speech, Signal Processing, vol. ASSP-27, pp. 4–12, 1979.

[168] A. Sonnenschein and B. W. Dickinson, "On a recent extrapolation procedure for band-limited signals," IEEE Trans. Circuits Syst., vol. CAS-29, pp. 116–117, 1982.

[169] A. K. Jain and S. Ranganath, "Extrapolation algorithms for discrete signals with application in spectral estimation," IEEE Trans. Acoust., Speech, Signal Processing, vol. ASSP-29, pp. 830–845, 1981.

[170] J. L. C. Snaz and T. H. Huang, "Discrete and continuous bandlimited signal extrapolation," IEEE Trans. Acoust., Speech, Signal Processing, vol. ASSP-31, pp. 1276–1285, 1983.

[171] W. Chujo and K. Kashiki, "Spherical array antenna using digital beamforming techniques for mobile satellite communications," Electron. Commun. Japan, (English trans. of Denshi Tsushin Gakkai Ronbunshi), vol. 75, pp. 76–86, 1992.

[172] R. Suzuki, Y. Matsumoto, R. Miura, and N. Hamamoto, "Mobile TDM/TDMA system with active array antenna," IEEE Global Telecommunications Conf. (GLOBECOM), Phoenix, AZ, 1991, pp. 1569–1573.

[173] H. Steyskal, "Digital beamforming antenna, an introduction," Microwave J., pp. 107–124, Jan. 1987.

[174] W. S. Youn and C. K. Un, "Eigenstructure method for robust array processing," Electron. Lett., vol. 26, pp. 678–680, 1990.

[175] A. M. Haimovich and Y. Bar-Ness, "An eigenanalysis interference canceller," IEEE Trans. Signal Processing, vol. 39, pp. 76–84, 1991.

[176] B. Friedlander, "A signal subspace method for adaptive interference cancellation," IEEE Trans. Acoust., Speech, Signal Processing, vol. 36, pp. 1835–1845, 1988.

[177] J. F. Yang and M. Kaveh, "Coherent signal-subspace transformation beam former," Proc. Inst. Elect. Eng., vol. 137, pt. F, pp. 267–275, 1990.

[178] B. D. Van Veen, "Eigenstructure based partially adaptive array design," IEEE Trans. Antennas Propagat., vol. 36, pp. 357–362, 1988.

[179] N. L. Owsley, "Sonar array processing," in Array Signal Processing, S. Haykin, Ed.. Englewood Cliffs, N.J.: Prentice-Hall, 1985, pp. 115–193.

[180] K. Nishimori, N. Kikuma, and N. Inagaki, "The differencial CMA adaptive array antenna using an eigen-beamspace system," IEICE Trans. Commun., vol. E78-B, pp. 1480–1488, Nov.1995.

[181] Alpesh U. Bhobe, Dr. Patrick L. Perini, "An overview of Smart Antenna Technology for Wireless communication", IEE 2000, pp 2-875 to 2-883

[182] M. Lisi, Alenia Spazio, " Antenna Technologies For Multimedia Mobile Satellite Communications ", IEE,11th International Conference on Antennas and Propagation,
Conference Publication No.480, pp 241 to 245

[183] Soren Anderson, Mille Millnert , Mats Viberg, Bo Wahlberg, " An Adaptive Array for Mobile Communication Systems", IEEE Transactions On Vechicular Technology, VOL.40,NO.1,February 1991, pp 230 to 236

[184] John S. Thompson, Peter M. Grant, Bernard Mulgrew, " Smart Antennas Arrays for CDMA Systems ", IEEE Personal Communications, October 1996, pp 16 to 22.

[185] Lal C. Godara, "Applications of Antenna Arrays to Mobile Communications, Part I: Performance Improvement, Feasibility, and System Considerations", Proceedings of the IEEE, VOL. 85, No. 7, July 1997, pp 1031-1060.

[186] Simon R. Saunders. "Antenna and propagation for wireless communication systems", John Wiley & Sons,LTD, 1999.

[187] LAL CHAND GODARA. 'Smart Antennas'. CRC press publications London, 2004.

[188] Bernard Widrow and Samuel D.Stearns. 'Adaptive Signal Processing'. Prentice Hall, 1985.

[189] Dr. B. Wardrop, " Digital Beamforming and Adaptive Techniques ", Marconi

[190] Tapan K.Sarkar, Michael C. Wicks, Magdalena Salazar-Palma, Robert J. Bonneau, " Smart Antennas ", IEEE Press, Wiley-Interscience. A John Wiley & sons, INC,Publications,2002

[191] Joseph C. Liberti, Jr. and Theodore S. Rappaport. 'Smart antennas for wireless communications: IS-95 and Third Generation CDMA Applications. Prentice Hall, 1999.

[192] B. Widrow, P.E. Mantey, L.J. Griffiths, B.B.Goode, " Adaptive Antenna Systems ", IEEE Proceedings. Vol.55, No.12 pp 2143 to 2159, December 1967.

[193] Tapan K. Sarkar, Norachet Sangruji, " An Adaptive Nulling System for a Narrow-Band Signal with a Look-Direction Constraint Utilizing the Conjugate Gradient Method ", IEEE Transactions on Antennas and Propagation, Vol. 37, No.7, July 1989,pp 940 to 944.

[194] Constantine A. Balanis, "Antenna Theory analysis & design",John Wiley & Sons,Inc, 2001.

[195] Roger F. Harrington, "Field Computation by Moment Methods ", Robert E. Krieger Publishing Company Malabar, Florida.

[196] Raviraj S. Adve, Tapan Kumar Sarkar, "Compensation for the Effects of Mutual Coupling on Direct Data Domain Adaptive Algorithms", IEEE Transactions on Antennas and propagation, VOL. 48, No.1, January 2000, pp 86 to 94.

[197] Inder J. Gupta, Aharon A. Ksienski, "Effect of mutual Coupling on the Performance of Adaptive Arrays", IEEE Transactions on Antennas and propagation, VOL. AP-31, No.5, September 1983, pp 785 to791

[198] Edward M. Friel, Krishna M. Pasala, " Effects of Mutual Coupling on the Performance of STAP Antenna Arrays ", IEEE Transactions On Aerospace And Electronic Systems, Vol. 36, No.2, April 2000, pp 518 to 527

[199] John L. Luzwick, Eugene C. Ngai, Arlon T. Adams, "Analysis of a Large Linear Antenna Array of Uniformly Spaced Thin Wire Dipoles Parallel to a

Perfectly Conducting Plane", IEEE Transactions on Antennas and Propagation, VOL. AP-30, No.2, March 1982, pp 230 to 234.

[200] Kun-Chou Lee Tah-Hsiung Chu, " A Circuit Model for Mutual Coupling Analysis of a finite Antenna Array", IEEE Transactions on Electromagnetic Compatibility, VOL. 38, No. 3, August 1996, pp 483 to 489.

[201] C. K. E. Lau, R. S. Adve, Tapan K. Sarkar, "Combined CDMA and Matrix Pencil Direction of Arrival Estimation", IEEE Proceedings on Vehicular Technology Conference 2002, pp 496 to 499.

[202] T. K. Sarkar, O. Pererira, " Using Matrix Pencil Method to Estimate the Parameters of a Sum of Complex Exponentials ," IEEE Antenna and Propagation Magazine, Vol.37, No.1, pp 48 to 55, February 1995.

[203] Feifei Gao, Alex B. Gershman, " A Generalized ESPRIT Approach to Direction –Of-Arrival", IEEE Signal Processing Letters, Vol.12, No.3, March 2005, pp 254 to 256

[204] Chien-Chung Yeh, Maw-Lin Leou, Donald R. Ucci, " Bearing Estimations with Mutual Coupling Present," IEEE Transactions on Antennas and Propagation, Vol. 37, No. 10, October 1989, pp 1332 to 1335.

[205] Simon C. Swales, Mark A. Beach, David J. Edwards, Joseph P. McGeehan, " The Performance Enhancement of Multibeam Adaptive Base-Station Antennas for Cellular Land Mobile Radio Systems", IEEE Transactions On Vehicular Technology, VOL.39, NO.1, February 1990, pp 56 to 67.

[206] Angeliki Alexiou, Martin Haardt, " Smart Antenna Technologies for Future Wireless Systems: Trends and Challenges", IEEE Communications Magazine, September 2004, pp 90 to 97

[207] Soren Anderson, Mille Millnert, Bo wahlberg, "An Adaptive Array for Mobile Communication Systems," IEEE Transactions on Vehicular Technology, Vol. 40, No.1, February 1991, pp 230 to 236.

[208] M. Lisi, Alenia Spazio, "Antenna Technologies for Multimedia Mobile Satellite Communications," 11th International Conference on Antennas and Propagation, 17-20 April 2001, Conference publication No. 480, IEE 2001.

[209] John S. Thompson, Peter M. Grant, and Bernard Mulgrew, "Smart Antenna Arrays for CDMA Systems," IEEE Personal Communications, October 1996, pp 16 to 22.

[210] Tapan K. Sarkar, Odilon Pereira, "Using the Matrix Pencil Method to Estimate the Parameters of a Sum of Complex Exponentials," IEEE Antennas and Propagation Magazine, Vol. 37, No. 1 ,February 1995, pp 48 to 55.

[211] John Farserotu, CSEM Ramjee Prasad, Aalborg University, "A Survey of Future Broadband Multimedia Satellite Systems, Issues and Trends", IEEE Communications Magazine, June 2000.pp 128-133.

[212] S.Drabowitch, A. Papiernik, H. Griffiths, J. Encinas and Bradford L. Smith, "Modern Antennas", Microwave and RF Technology Series 12, Chapman & Hall, 1998.

[213] L. J. Horowitz, H. Blatt, W. G. Brodsky, and K. D. Senne, "Controlling adaptive antenna arrays with the sample matrix inversion algorithm," IEEE Trans. Aerosp. Electron. Syst., vol. AES-15, pp. 840–847, 1979.

[214] R. Schreiber, "Implementation of adaptive array algorithms," IEEE Trans. Acoust., Speech, Signal Processing, vol. ASSP-34, pp. 1038–1045, 1986.

[215] E. Lindskog, "Making SMI-beamforming insensitive to the sampling timing for GSM signals," in Proc. IEEE Int. Symp. Personal, Indoor and Mobile Radio Communications,Toronto, Canada, 1995, pp. 664–668.

[216] R. G. Vaughan, "On optimum combining at the mobile," IEEE Trans. Veh. Technol., vol. 37, pp. 181–188, 1988.

[217] H. Hashemi, "The indoor radio propagation channels," Proc. IEEE, vol. 81, pp. 943–968, 1993.

[218] C. Passerini, M. Missiroli, G. Riva, and M. Frullone, "Adaptive antenna arrays for reducing the delay spread in indoor radio channels," Electron. Lett., vol. 32, pp. 280–281, 1996.

[219] L. J. Griffiths, "A simple adaptive algorithm for real-time processing in antenna arrays," Proc. IEEE, vol. 57, pp. 1696–1704, 1969.

[220] B. Widrow and J. M. McCool, "A comparison of adaptive algorithms based on the methods of steepest descent and random search," IEEE Trans. Antennas Propagat., vol. AP-24, pp. 615–637, 1976.

[221] R. A. Iltis and L. B. Milstein, "An approximate statistical analysis of the Widrow LMS algorithm with application to narrow-band interference rejection," IEEE Trans. Commun., vol. COM-33, pp. 121–130, 1985.

[222] P. M. Clarkson and P. R. White, "Simplified analysis of the LMS adaptive filter using a transfer function approximation," IEEE Trans. Acoust., Speech, Signal Processing, vol. ASSP-35, pp. 987–993, 1987.

[223] W. A. Gardner, "Comments on convergence analysis of LMS filters with uncorrelated data," IEEE Trans. Acoust., Speech, Signal Processing, vol. ASSP-34, pp. 378–379, 1986.

[224] J. B. Foley and F. M. Boland, "A note on the convergence analysis of LMS adaptive filters with Gaussian data," IEEE Trans. Acoust., Speech, Signal Processing, vol. 36, pp. 1087–1089, 1988.

[225] V. Solo, "The limiting behavior of LMS," IEEE Trans. Acoust., Speech, Signal Processing, vol. 37, pp. 1909–1922, 1989.

[226] A. Feuer and E. Weinstein, "Convergence analysis of LMS filters with uncorrelated Gaussian data," IEEE Trans. Acoust., Speech, Signal Processing, vol. ASSP-33, pp. 222–229, 1985.

[227] S. Jaggi and A. B. Martinez, "Upper and lower bounds of the misadjustment in the LMS algorithm," IEEE Trans. Acoust., Speech, Signal Processing, vol. 38, pp. 164–166, 1990.

[228] F. B. Boland and J. B. Foley, "Stochastic convergence of the LMS algorithm in adaptive systems," Signal Process., vol. 13, pp. 339–352, 1987.

[229] B. Widrow, J. McCool, M. G. Larimore, and C. R. Johnston Jr., "Stationary and nonstationary learning characteristics of the LMS adaptive filter," in Aspects of Signal Processing, part 1, G. Tacconi, Ed. Boston, MA: Reidel, 1976, pp. 355–393; also
in Proc. IEEE, vol. 64, pp. 1151–1162, 1976.

[230] L. H. Horowitz and K. D. Senne, "Performance advantage of complex LMS for controlling narrow-band adaptive arrays," IEEE Trans. Circuits Syst., vol. CAS-28, pp. 562–576, 1981.

[231] V. Solo, "The error variance of LMS with time-varying weights," IEEE Trans. Signal Processing, vol. 40, pp. 803–813, 1992.

[232] M. Nagatsuka, N. Ishii, R. Kohno, and H. Imai, "Adaptive array antenna based on spatial spectral estimation using maximum entropy method," IEICE Trans. Commun., vol. E77-B, pp. 624–633, 1994.

[233] J. S. Soo and K. K. Pang, "A multiple size frequency domain adaptive filter," IEEE Trans. Signal Processing, vol. 39, pp. 115–121, 1991.

[234] Z. Pritzker and A. Feuer, "Variable length stochastic gradient algorithm," IEEE Trans. Signal Processing, vol. 39, pp. 997–1001, 1991.

[235] F. F. Yassa, "Optimality in the choice of the convergence factor for gradient based adaptive algorithms," IEEE Trans. Acoust., Speech, Signal Processing, vol. ASSP-35, pp. 48–59, 1987.

[236] J. B. Evans, P. Xue, and B. Liu, "Analysis and implementation of variable step size adaptive algorithms," IEEE Trans. Signal Processing, vol. 41, pp. 2517–2535, 1993.

[237] R. H. Kwong and E. W. Johnston, "A variable step size LMS algorithm," IEEE Trans. Signal Processing, vol. 40, pp. 1633–1642, 1992.

[238] C. C. Ko, "A fast adaptive null-steering algorithm based on output power measurements," IEEE Trans. Aerosp. Electron. Syst., vol. 29, pp. 717–725, 1993.

[239] C. C. Ko, G. Balabshaskar, and R. Bachl, "Unbiased source estimation with an adaptive null steering algorithm," Signal Process., vol. 31, pp. 283–300, 1993.

[240] J. Benesty and P. Duhamel, "A fast exact least mean square adaptive algorithm," IEEE Trans. Signal Processing, vol. 40, pp. 2904–2920, 1992.

[241] A. Feuer and R. Cristi, "On the steady state performance of frequency domain LMS algorithms," IEEE Trans. Signal Processing, vol. 41, pp. 419–423, 1993.

[242] V. J. Mathews and S. H. Cho, "Improved convergence analysis of stochastic gradient adaptive filters using the sign algorithm," IEEE Trans. Acoust., Speech, Signal Processing, vol. ASSP-35, pp. 450–454, 1987.

[243] N. J. Bershad and L. Z. Qu, "LMS adaptation with correlated data—A scalar example," IEEE Trans. Acoust., Speech, Signal Processing, vol. ASSP-32, pp. 695–700, 1984.

[244] N. J. Bershad and Y. H. Chang, "Time correlation statistics of the LMS adaptive algorithm weights," IEEE Trans. Acoust., Speech, Signal Processing, vol. ASSP-33, pp. 309–312, 1985.

[245] E. Eweda, "Analysis and design of a signed regressor LMS algorithm for stationary and non stationary adaptive filtering with correlated Gaussian data," IEEE Trans. Circuits Syst., vol. 37, pp. 1367–1374, 1990.

[246] S. Kaczmarz, "Angenaherte Auflosung von Systemen linearen gleichungen," Bull. Int. Acad. Pol. Sci. Lett., 1937.

[247] Y. Z. Tsypkin, Foundation of the Theory of Searching Systems New York: Academic, 1973.

[248] J. I. Nagumo and A. Noda, "A learning method for system identification," IEEE Trans. Automat. Contr., vol. AC-12, pp. 282–287, 1967.

[249] R. Nitzberg, "Application of the normalized LMS algorithm to MSLC," IEEE Trans. Aerosp. Electron. Syst., vol. AES-21, pp. 79–91, 1985.

[250] , "Normalized LMS algorithm degradation due to estimation noise," IEEE Trans. Aerosp. Electron. Syst., vol. AES-22, pp. 740–750, 1986.

[251] N. J. Bershad, "Analysis of the normalized LMS algorithm with Gaussian inputs," IEEE Trans. Acoust., Speech, Signal Processing, vol. ASSP-34, pp. 793–806, 1986.

[252] D. T. M. Slock, "On the convergence behavior of the LMS and the normalized LMS algorithms," IEEE Trans. Signal Processing, vol. 41, pp. 2811–2825, 1993.

[253] M. Rupp, "The behavior of LMS and NLMS algorithms in the presence of spherically invariant processes," IEEE Trans. Signal Processing, vol. 41, pp. 1149–1160, 1993.

[254] M. Barrett and R. Arnott, "Adaptive antennas for mobile communications," Electron. Commun. Eng. J., vol. 6, pp. 203–214, 1994.

XIII

[255] A. Cantoni, "Application of orthogonal perturbation sequences to adaptive beamforming," IEEE Trans. Antennas Propagat., vol. AP-28, pp. 191–202, 1980.

[256] L. C. Godara and A. Cantoni, "Analysis of the performance of adaptive beamforming using perturbation sequences," IEEE Trans. Antennas Propagat., vol. AP-31, pp. 268–279, 1983.

[257] , "Analysis of constrained LMS algorithm with application to adaptive beamforming using perturbation sequences," IEEE Trans. Antennas Propagat., vol. AP-34, pp. 368–379, 1986.

[258] J. L. Moschner, "Adaptive filters with clipped input data," Information Systems Laboratory, Stanford University, CA, Tech. Rep. 6796-1, 1970.

[259] L. C. Godara, "Constrained beamforming and adaptive algorithms," in Handbook of Statistics, vol. 10, N. K. Bose and C. R. Rao, Eds. Amsterdam, The Netherlands: Elsevier, 1993.

[260] , "Performance analysis of structured gradient algorithm," IEEE Trans. Antennas Propagat., vol. 38, 1078–1083, 1990.

[261] L. C. Godara and D. A. Gray, "A structure gradient algorithm for adaptive beamforming," J. Acoust. Soc. Amer., vol. 86, pp. 1040–1046, 1989.

[262] L. C. Godara, "Improved LMS algorithm for adaptive beamforming," IEEE Trans. Antennas Propagat., vol. 38, pp. 1631–1635, 1990.

[263] T. Ohgane, N . Matsuzawa, T. Shimura, M. Mizuno, and H. Sasaoka, "BER performance of CMA adaptive array for high-speed GMSK mobile communication— A description of measurements in central Tokyo," IEEE Trans. Veh. Technol., vol. 42, pp. 484–490, 1993.

[264] R. M. Davis, D. C. Farden, and P. J. S. Sher, "A coherent perturbation algorithm," IEEE Trans. Antennas Propagat., vol. AP-34, pp. 380–387, 1986.

[265] B. Farhang-Boroujeny and L. F. Turner, "Fast converging stochastic gradient algorithm," Proc. Inst. Elect. Eng., vol. 128, pt. F, pp. 271–274, 1981.

[266] S. T. Alexander, "Transient weight misadjustment properties for the finite precision LMS algorithm," IEEE Trans. Acoust., Speech, Signal Processing, vol. ASSP-35, pp. 1250–1258, 1987.

[267] Y. H. Chang, C. K. Tzou, and N. J. Bershad, "Postsmoothing for the LMS algorithm and a fixed point round-off error analysis,"
IEEE Trans. Signal Processing, vol. 39, pp. 959–962, 1991.

[268] P. W. Wong, "Quantization and roundoff noises in fixed-point FIR digital filters," IEEE Trans. Signal Processing, vol. 39, pp. 1552–1563, 1991.

[269] R. D. Gitlin, J. E. Mago, and M. G. Taylor, "On the design of gradient algorithms for digitally implemented adaptive filters," IEEE Trans. Circuit Theory, vol. CT-20, pp. 125–136, 1973.

[270] C. Caraiscos and B. Liu, "A round-off error analysis of the LMS adaptive algorithm," IEEE Trans. Acoust., Speech, Signal Processing, vol. ASSP-32, pp. 34–41, 1984.

[271] R. D. Girlin, H. C. Meadors, and S. B. Weinstein, "The tap leakage algorithm: An algorithm for stable operations of a digitally implemented, fractionally spaced equalizer," Bell Syst. Tech. J., vol. 61, pp. 1817–1839, 1982.

[272] J. M. Cioffi and J. J. Werner, "Effect of biases on digitally implemented data driven echo canceller," AT&T Tech. J., vol. 64, pp. 115–138, 1985.

[273] A. V. Oppenheim and R. W. Schafer, Digital Signal Processing. Englewood Cliffs, NJ: Prentice-Hall, 1975.

[274] B. Widrow, J. McCool, and M. Ball, "The complex LMS algorithm," Proc. IEEE, vol. 63, pp. 719–720, 1975.

[275] A. Papoulis, Probability, Random Variables and Stochastic Processes. New York: McGraw-Hill, 1965.

[276] E. Eweda and O. Macchi, "Convergence of the RLS and LMS adaptive filters," IEEE Trans. Circuits Syst., vol. CAS-34, pp. 799–803, 1987.

[277] P. Fabre and C. Gueguen, "Improvement of the fast recursive least-squares algorithms via normalization: A comparative study," IEEE Trans. Acoust., Speech, Signal Processing, vol. ASSP-34, pp. 296–308, 1986.

[278] P. E. Mantey and L. J. Griffiths, "Iterative least-squares algorithm for signal extraction," in 2nd Int. Hawaii Conf. System Science, Honolulu, HI, 1969, pp. 767–770.

[279] E. Eleftheriou and D. D. Falconer, "Tracking properties and steady state performance of RLS adaptive filter algorithms," IEEE Trans. Acoust., Speech, Signal Processing, vol. ASSP-34, pp. 1097–1110, 1986.

[280] M. S. Mueller, "Least squares algorithms for adaptive equalizers," Bell Syst. Tech. J., pp. 1905–1925, 1981.

[281] J. M. Cioffi and T. Kailath, "Fast recursive-least-square, transversal filters for adaptive filtering," IEEE Trans. Acoust., Speech, Signal Processing, vol. ASSP-32, pp. 998–1005, 1984.

[282] T. Murali and B. V. Rao, "A class of recursive maximumlikelihood algorithms," Proc. IEEE, vol. 73, pp. 1336–1339, 1985.

[283] G. V. Moustakids and S. Theodoridis, "Fast Newton transversal filters—A new class of adaptive estimation algorithms," IEEE Trans. Signal Processing, vol. 39, pp. 2184–2193, 1991.

[284] S. Qiao, "Fast adaptive RLS algorithms: A generalized inverse approach and analysis," IEEE Trans. Signal Processing, vol. 39, pp. 1455–1459, 1991.

[285] D. T. M. Slock and T. Kailath, "Numerically stable fast transversal filters for recursive least squares adaptive filtering," IEEE Trans. Signal Processing, vol. 39, pp. 92–114, 1991.

[286] W. A. Gardner and W. A. Brown III, "A new algorithm for adaptive arrays," IEEE Trans. Acoust., Speech, Signal Processing, vol. ASSP-35, pp. 1314–1319, 1987.

[287] J. Fernandez, I. R. Corden, and M. Barrett, "Adaptive array algorithms for optimal combining in digital mobile communication systems," in Proc. Inst. Elect. Eng. 8th Int. Conf. Antennas and Propagation, Edinburgh, Scotland, 1993, pp. 983–986.

[288] Y. Wang and J. R. Cruz, "Adaptive antenna arrays for the reverse link of CDMA cellular communication systems," Electron. Lett., vol. 30, pp. 1017–1018, 1994.

[289] D. N. Godard, "Self-recovering equalization and carrier tracking in two-dimensional data communication systems," IEEE Trans. Commun., vol. COM-28, pp. 1867–1875, 1980.

[290] J. R. Treichler and B. G. Agee, "A new approach to multipath correction of constant modulus signals," IEEE Trans. Acoust., Speech, Signal Processing, vol. ASSP-31, pp. 459–472, 1983.

[291] J. J. Shynk and C. K. Chan, "Performance surfaces of the constant modulus algorithm based on a conditional Gaussian model," IEEE Trans. Signal Processing, vol. 41, pp. 1965–1969, 1993.

[292] T. Ohgane, "Characteristics of CMA adaptive array for selective fading compensation in digital land mobile radio communications," Electron. Commun. Jpn., vol. 74, pp. 43–53, 1991.

[293] T. Ohgane, T. Shimura, N. Matsuzawa, and H. Sasaoka, "An implementation of a CMA adaptive array for high speed GMSK transmission in mobile communications," IEEE Trans. Veh. Technol., vol. 42, pp. 282–288, 1993.

[294] I. Parra, G. Xu, and H. Liu, "Least squares projective constant modulus approach," in Proc. IEEE Int. Symp. Personal, Indoor and Mobile Radio Communications, Toronto, Canada, 1995, pp. 673–676.

[295] M. Hestenes and E. Stiefel, "Method of conjugate gradients for solving linear systems," J. Res. Natl. Bur. Stand., vol. 49, pp. 409–436, 1952.

[296] J. W. Daniel, "The conjugate gradient method for linear and nonlinear operator equations," SIAM J. Numer. Anal., vol. 4, pp. 10–26, 1967.

[297] T. Sarkar, K. R. Siarkiewicz, and R. F. Stratton, "Survey of numerical methods for solutions of large systems of linear equations for electromagnetic field problems," IEEE Trans. Antennas Propagat., vol. AP-29, pp. 847–856, 1981.

[298] S. Choi and D. H. Kim, "Adaptive antenna array utilizing the conjugate gradient method for compensation of multipath fading in a land mobile communication," in Proc. IEEE 42nd Vehicular Technology Conf., Denver, CO, 1992, pp. 33–36.

[299] S. Choi, Application of the Conjugate Gradient Method for Optimum Array Processing, vol. V. Amsterdam, The Netherlands: Elsevier, 1991, ch. 16.

[300] L. C. Godara, "Limitations and capabilities of directions-ofarrival estimation techniques using an array of antennas: A mobile communications perspective," presented at the IEEE Int. Symp. Phased Array Systems and Technology, Boston, MA, 1996.

[301] R. T. Lacoss, "Data adaptive spectral analysis method," Geophysics, vol. 36, pp. 661–675, 1971.

[302] A. H. Nuttall, G. C. Carter, and E. M. Montaron, "Estimation of two-dimensional spectrum of the space-time noise field for a sparse line array," J. Acoust. Soc. Amer., vol. 55, pp. 1034–1041, 1974.

[303] D. H. Johnson, "The application of spectral estimation methods to bearing estimation problems," Proc. IEEE, vol. 70, pp. 1018–1028, 1982.

[304] R. A. Wagstaff and J. L. Berrou, "A fast and simple nonlinear technique for high-resolution beamforming and spectral analysis," J. Acoust. Soc. Amer., vol. 75, pp. 1133–1141, 1984.

[305] Q. T. Zhang, "A statistical resolution theory of the beamformerbased spatial spectrum for determining the directions of signals in white noise," IEEE Trans. Signal Processing, vol. 43, pp. 1867–1873, 1995.

[306] M. S. Bartlett, An Introduction to Stochastic Process. New York: Cambridge Univ. Press, 1956.

[307] V. A. N. Barroso, M. J. Rendas, and J. P Gomes, "Impact of array processing techniques on the design of mobile communication systems," in Proc. IEEE 7th Mediterranean Electrotechnical, Antalya, Turkey, 1994, pp. 1291–1294.

[308] M. I. Miller and D. R. Fuhrmann, "Maximum likelihood narrow-band direction finding and the EM algorithm," IEEE Trans. Acoust., Speech, Signal Processing, vol. 38, pp. 1560–1577, 1990.

[309] J. Makhoul, "Linear prediction: A tutorial review," Proc. IEEE, vol. 63, pp. 561–580, 1975.

[310] S. B. Kesler, S. Boodaghians, and J. Kesler, "Resolving uncorrelated and correlated sources by linear prediction," IEEE Trans. Antennas Propagat., vol. AP-33, pp. 1221–1227, 1985.

[311] J. P. Burg, "Maximum entropy spectral analysis," presented at the 37th Annu. Meeting, Society Exploration Geophysics, Oklahoma City, OK, 1967.

[312] J. H. McClellan and S. W. Lang, "Duality for multidimensional MEM spectral analysis," Proc. Inst. Elect. Eng., vol. 130, pt. F, pp. 230–235, 1983.

[313] D. P. Skinner, S. M. Hedlicka, and A. D. Mathews, "Maximum entropy array processing," J. Acoust. Soc. Amer., vol. 66, pp. 488–493, 1979.

[314] T. Thorvaldsen, "Maximum entropy spectral analysis in antenna spatial filtering," IEEE Trans. Antennas Propagat., vol. AP-28, no. 99, pp. 556–562, 1980.

[315] J. H. McClellan, "Multidimensional spectral estimation," Proc. IEEE, vol. 70, pp. 1029–1039, 1982.

[316] S. W. Lang and J. H. McClellan, "Spectral estimation for sensor arrays," IEEE Trans. Acoust., Speech, Signal Processing, vol. ASSP-31, pp. 349–358, 1983.

[317] D. R. Farrier, "Maximum entropy processing of band-limited spectra, Part 1: Noise free case," Proc. Inst. Elect. Eng., vol. 132, pt. F, pp. 491–504, 1985.

[318] W. S. Ligget, "Passive sonar: Fitting models to multiple timeseries," in NATO ASI Signal Processing, J. W. R. Griffiths et al., Eds. New York: Academic, 1973, pp. 327–345.

[319] F. C. Schweppe, "Sensor array data processing for multiple signal sources," IEEE Trans. Inform. Theory, vol. IT-14, pp. 294–305, 1968.

[320] I. Ziskind and M. Wax, "Maximum likelihood localization of multiple sources by alternating projection," IEEE Trans. Acoust., Speech, Signal Processing, vol. ASSP-36, pp. 1553–1560, 1988.

[321] P. Stoica and K. C. Sharman, "Maximum likelihood methods for direction of arrival estimation," IEEE Trans. Acoust., Speech, Signal Processing, vol. ASSP-38, pp. 1132–1143, 1990.

[322] S. K. Oh and C. K. Un, "Simple computational methods of the AP algorithm for maximum likelihood localization of multiple radiating sources," IEEE Trans. Signal Processing, vol. 40, pp. 2848–2854, 1992.

[323] H. Lee and R. Stovall, "Maximum likelihood methods for determining the direction of arrival for a single electromagnetic source with unknown polarization," IEEE Trans. Signal Processing, vol. 42, pp. 474–479, 1994.

[324] Q. Wu, K. M. Wong, and J. P. Reilly, "Maximum likelihood direction finding in unknown noise environments," IEEE Trans. Signal Processing, vol. 42, pp. 980–983, 1994.

[325] J. Sheinvald, M. Wax, and A. J. Weiss, "On maximumlikelihood localization of coherent signals," IEEE Trans. Signal Processing, vol. 44, pp. 2475–2482, 1996.

[326] S. Haykin, "Radar array processing for angle of arrival estimation," in Array Signal Processing, S. Haykin, Ed. Englewood Cliffs, NJ: Prentice-Hall, 1985.

[327] M. Wax and T. Kailath, "Optimum localization of multiple sources by passive arrays," IEEE Trans. Acoust., Speech, Signal Processing, vol. ASSP-31, pp. 1210–11221, 1983.

[328] A. D. Dempster, N. M. Laird, and D. B. Rubin, "Maximum likelihood from incomplete data via the EM algorithm," J. Roy Statist. Soc., vol. 13–19, pp. 1–37, 1977.

[329] M. J. Hinich, "Frequency-wave number array processing," J. Acoust. Soc. Amer., vol. 69, pp. 732–737, 1981.

[330] T. J. Abatzoglou, "A fast maximum likelihood algorithm for frequency estimation of a sinusoid based on Newton's method," IEEE Trans. Acoust., Speech, Signal Processing, vol. ASSP-33, pp. 77–89, 1985.

[331] T. Wigren and A. Eriksson, "Accuracy aspects of DOA and angular velocity estimation in sensor array processing," IEEE Signal Processing Lett., vol. 2, pp. 60–62, 1995.

[332] A. Zeira and B. Friedlander, "On the performance of direction finding with time varying arrays," Signal Process., vol. 43, pp. 133–147, 1995.

[333] D. W. Tufts and C. D. Melissinos, "Simple, effective computation of principal eigenvectors and their eigenvalues and application to high-resolution estimation of frequencies," IEEE Trans. Acoust., Speech, Signal Processing, vol. ASSP-34, pp. 1046–1053, 1986.

[334] V. F. Pisarenko, "The retrieval of harmonics from a covariance function," Geophys. J. R. Astron. Soc., vol. 33, pp. 347–366, 1973.

[335] M. Wax, T. J. Shan, and T. Kailath, "Spatio-temporal spectral analysis by eigenstructure methods," IEEE Trans. Acoust., Speech, Signal Processing, vol. ASSP-32, pp. 817–827, 1984.

[336] S. S. Reddi, "Multiple source location—A digital approach," IEEE Trans. Aerosp. Electron. Syst., vol. AES-15, pp. 95–105, 1979.

[337] A. Cantoni and L. C. Godara, "Resolving the directions of sources in a correlated field incident on an array," J. Acoust. Soc. Amer., vol. 67, pp. 1247–1255, 1980.

[338] D. J. Bordelon, "Complementarity of the Reddi method of source direction estimation with those of Pisarenko and Cantoni and Godara, I," J. Acoust. Soc. Amer., vol. 69, pp. 1355–1359, 1981.

[339] D. H. Johnson and S. R. DeGraff, "Improving the resolution of bearing in passive sonar arrays by eigenvalue analysis," IEEE Trans. Acoust., Speech, Signal Processing, vol. ASSP-29, pp. 401–413, 1982.

[340] T. P. Bronez and J. A. Cadzow, "An algebraic approach to super resolution array processing," IEEE Trans. Aerosp. Electron. Syst., AES-19, pp. 123–133, 1983.

[341] V. U. Reddy, B. Egardt, and T. Kailath, "Least-squares type algorithm for adaptive implementation of Pisarenko's harmonic retrieval method," IEEE Trans. Acoust., Speech, Signal Processing, vol. ASSP-30, pp. 399–405, 1982.

[342] J. R. Yang and M. Kaveh, "Adaptive eigensubspace algorithms for direction or frequency estimation and tracking," IEEE Trans. Acoust., Speech, Signal Processing, vol. ASSP-36, pp. 241–251, 1988.

[343] M. G. Larimore, "Adaptive convergence of spectral estimation based on Pisarenko harmonic retrieval," IEEE Trans. Acoust., Speech, Signal Processing, vol. ASSP-31, pp. 955–962, 1983.

[344] H. Ouibrahim, "Prony, Pisarenko and the matrix pencil: A unified presentation," IEEE Trans. Acoust., Speech, Signal Processing, vol. ASSP-37, pp. 133–134, 1989.

[345] H. Ouibrahim, D. D. Weiner, and T. K. Sarkar, "A generalized approach to direction finding," IEEE Trans. Acoust., Speech, Signal Processing, vol. 36, pp. 610–613, 1988.

[346] A. Paulraj and T. Kailath, "Eigenstructure methods for direction of arrival estimation in the presence of unknown noise field," IEEE Trans. Acoust., Speech, Signal Processing, vol. ASSP-34, pp. 13–20, 1986.

[347] M.Wax, "Detection and localization of multiple sources in noise with unknown covariance," IEEE Trans. Signal Processing, vol. 40, pp. 245–249, 1992.

[348] A. J. Weiss, A. S. Willsky, and B. C. Levy, "Eigenstructure approach for array processing with unknown intensity coefficients," IEEE Trans. Acoust., Speech, Signal Processing, vol. ASSP-36, pp. 1613–1617, 1988.

[349] G. Bienvenu, "Influence of the spatial coherence of the back- ground noise on high resolution passive methods," in Proc. ICASSP, Washington, D.C., 1979, pp. 306–309.

CHAPTER 1

Smart Antennas for Satellite based Mobile Communication: Benefit and Challenges

1. Introduction

Over the last few years the demand for service provision via the wireless communication bearer has risen beyond all expectations. If the extraordinary fact that worldwide some half a billion subscribers to mobile networks are predicted by the year 2005 is put in the context of third generation system requirements (UMTS, IMT2000) then the most demanding technological challenge emerges that is the need to increase the spectrum efficiency of wireless networks. While great effort in current (second) generation wireless communication systems has been directed towards the development of modulation, coding and protocols, antenna-related technology has received significantly less attention up to now. To achieve the ambitious requirements introduced for future wireless systems, new 'intelligent' or 'self-configured' and highly efficient systems will most certainly be required. In the pursuit of schemes that will solve these problems, attention has recently turned to spatial filtering methods using advanced antenna techniques: adaptive or smart antennas. Filtering in the space domain can separate spectrally and temporally overlapping signals from multiple mobile units, and hence the spatial dimension can be exploited as a hybrid multiple access technique complementing FDMA, TDMA and CDMA.

1.1 Requirements of Antenna Arrays and Smart Antennas in Satellite based mobile communication

The demand for wireless mobile communications services is growing at an explosive rate, with the anticipation that communication to a mobile device anywhere on the globe at all times will be available in the near future. Satellite personal communications systems have been able to provide communication services to vast regions of the earth that lack adequate infrastructure for commercial telephony services. Satellite personal communication services are comparable to moving cellular technology into space. In satellite as well as cellular mobile communications, the limiting factors are shadowing effects, power control, hand-off, multipath fading, capacity and cost. Most existing cellular systems use simple antennas, transmitting in a fixed direction or all directions (omni directional). This causes a lot of interference between the subscribers. On control channels interference will lead to missed or blocked calls due to errors in digital signaling. With rapid increase in cellular subscription, capacity will be an issue in future communication systems. Satellite communication is meant to provide multimedia services, data communication, and high speed internet anywhere on the earth at any time. For high speed application higher bandwidth is required. To accommodate millions of simultaneous users it is necessary to implement frequency reuse scheme because available frequency spectrum for such applications is limited. Frequency reuse scheme at the satellite implies the generation of multiple beams. Now multiple beams can be easily generated using antenna arrays. However in case of Leo satellite constellations, because of satellite motions, and earths rotation frequent hand-offs are necessary therefore to reduce, the number of hand-offs beam steerable antennas are required. An application of antenna arrays has been suggested in recent years for mobile communications systems to overcome the problem of limited channel bandwidth, thereby satisfying an ever growing demand for a large number of mobiles on communications channels. It has been shown by many studies that when an array is appropriately used in a mobile communications system may be cellular mobile or personal satellite communications, it helps in improving the system performance by increasing channel capacity and spectrum efficiency, extending range coverage, tailoring beam shape, steering multiple beams to track many mobiles. It also reduces multipath fading, co-channel interferences, BER (Bit Error Rate) and outage probability. It has been urged that adaptive antennas (Smart Antenna System) and algorithms to control them are vital to a high capacity communications system development. An adaptive array of antennas may be used in a variety of ways to improve the performance of a communications system. Perhaps most important is its capability to cancel co-channel interferences. An adaptive array antenna system works on the premise that the desired signal and unwanted co-channel interference arrive from different directions. The beam pattern of the array is adjusted to suit the requirements by combining signals from different antennas with appropriate weighting. The scheme needs to differentiate the desired signal from the co-channel interferences and normally requires either the knowledge of a reference signal, a training signal or the direction of the desired signal source to achieve its desired objectives. There exists a range of schemes to estimate the direction of sources with conflicting demands of accuracy and processing power. Similarly, there are many methods and algorithms to update the array weights, each with its speed of convergence and required processing time. Algorithms also exist that exploit properties of signals to eliminate the need of training signals in some circumstances.

There has been a wide range of research covering development of antennas suitable for mobile communications systems, and many experimental results have been reported to show the system requirements and feasibility. It is anticipated that a future mobile communications system would consist of a hand-held terminal the size of a wristwatch capable of steering a beams, toward a satellite. The system would also consist of many radiating elements fabricated by micro-strip technology, each with its own phase-shifting network, power amplifier, and so on along with other required processors manufactured by the microwave monolithic integrated circuits technology. And all this is expected to be available at an affordable price. Spatial diversity is one technique that can be implemented using adaptive antenna array. Space diversity (sometimes known as antenna diversity) is widely used in wireless systems. Recent trends in mobile communication system have shown that adaptive (smart) antenna arrays have tremendous potential for increasing the capacity of mobile communication by reducing co-channel interference multipath and noise.

1.2 Performance improvement in satellite and cellular mobile communication using an adaptive antenna arrays

An antenna array is able to improve the performance of a mobile communication system in a number of ways. It provides the capability to reduce co-channel interferences and multipath fading, resulting in better quality of services, such as reduced BER and outage probability. Its capability to form multiple beams could be exploited to serve many users in parallel, resulting in an increased spectral efficiency. Its ability to adapt beam shapes to suit traffic conditions is useful in reducing the handoff rate. The advantages of adaptive antennas in mobile communication systems

1) Reduction in delay spread and multipath fading

Delay spread is caused by multipath propagation where a desired signal arriving from different directions gets delayed due to the different travel distances involved. An array with the capability to form beams in certain directions and nulls in others is able to cancel some of these delayed arrivals in two ways. First, in the transmit mode, it focuses energy in the required direction, which helps to reduce multipath reflections causing a reduction in the delay spread. Second, in the receive mode, an antenna array provides compensation in multipath fading by diversity combining, by adding the signals belonging to different clusters after compensating for delays, and by canceling delayed signals arriving from directions other than that of the main signal point

a) *Use of diversity combining*

Diversity combining achieves a reduction in fading by increasing the signal level based upon the level of signal strength at different antennas, whereas in multipath cancellation methods, it is achieved by adjusting the beam pattern to accommodate nulls in direction of late arrivals, assuming them to be as interferences. For the latter case, a beam is pointed in the direction of the direct path or a path along which a major component of the signal arrives, causing the reduction in the energy received from other directions and thus reducing the components of multipath signal contributing to the receiver.

b) *Combining delayed arrivals*:

A radio wave originating from a source arrives at a distant point in clusters after getting scattered and reflected from objects along the way. This is particularly true in scenarios with large buildings and hills where delayed arrivals are well separated. One could use these clustered signals constructively by grouping them as per their delays compared to a signal available from the shortest path. Individual paths of these delayed signals may be resolved by exploiting their temporal or spatial structure. The resolution of paths using temporal structures depends upon the bandwidth of the signal compared to the coherence multipath. It is argued that due to the distributed nature of sources, synthesis of antenna pattern with nulls toward interferences is not practical. It suggests that use of optimal combining of a desired signal arriving from the various paths, using a reference signal transmitted from the base in the form of a pilot signal for a spread spectrum system.

2) Reduction in co-channel interference

An antenna array has the property of spatial filtering, which may be exploited in transmitting as well as in receiving modes to reduce co-channel interferences. In the transmitting mode, it can be used to focus radiated energy by forming a directive beam in a small area where a receiver is likely to be. This means there is less interference in other directions where the beam is not pointing.

Co-channel interference in transmit mode could be further reduced by forming specialized beams with nulls in directions of other receivers. In receiving mode, when the precise direction of the signal is known, interference cancellation may be achieved by solving a constrained beam forming problem, whereas when this is not the case, it may be achieved by using a reference signal. A comparison of the two schemes for a satellite to satellite communications scenario when an approximate direction of the signal is known indicates that the constrained scheme cancels the interference faster than the reference generation scheme.

3) Spectrum efficiency and capacity improvement

Spectrum efficiency refers to the amount of traffic a given system with certain spectrum allocation could handle. An increase in the number of users of the mobile communications system without a loss of performance causes the spectrum efficiency to increase. Channel capacity refers to the maximum data rate a channel of given bandwidth could sustain. An improved channel capacity leads to more users of a specified data rate, implying better spectrum efficiency.

4) BER improvement

A consequence of a reduction in co-channel interference and multipath fading by using an adaptive array in a mobile communications system to improve the communications quality is a reduction in BER (Bit Error Rate) and SER (Symbol Error Rate) for a given SNR or a reduction in required SNR for a given BER.

5) Reduction in outage probability

Outage probability is the probability of a channel's being inoperative due to increased error rate in the received data. It may be caused by co-channel interference in a mobile communication system. Using an array helps to reduce the outage probability by decreasing co-channel interference. It decreases as the number of beams used increases in a multibeam adaptive antenna system.

6) Increase in transmission efficiency

Electronically steerable antennas are directive compared to fixed omnidirectional antennas, that is, they have high gains in the direction where the beam is pointing. This fact may be useful in extending the range, or may be used to reduce the transmitted power of the mobiles. This follows from the fact that by using a highly directive antenna, a satellite may be able to pick a weaker signal within the cell than by using an omnidirectional antenna. This in turn means that the mobile has to transmit less power and its battery life will last longer, or it would be able to use a smaller battery, resulting in a smaller size and weight.

7) Dynamic channel assignment

In mobile communications, channels are generally assigned in a fixed manner depending upon the position of a mobile and the available channels in the cell where the mobile is positioned. As a mobile crosses the cell boundary, a new channel is assigned. In this arrangement, the number of channels in a cell is normally fixed. The use of adaptive array provides an opportunity to change the cell boundary and thus to allocate the number of channels in each cell as the demand changes due to changed traffic situation. This provides the means whereby a mobile or group of mobiles may be tracked as it moves and the cell boundary may be adjusted to suit this group.

8) Reduction in Handoff rate

When the number of mobiles in a cell exceeds its capacity, cell splitting is used to create new cells having new frequency assignment. A consequence of this is an increased handoff due to reduced cell size. This may reduced using adaptive array antennas. Instead of cell splitting, the capacity is increased by creating independent beams using more antennas. Each beam is adjusted as the mobile locations change. The beam follows a cluster of mobiles or a single mobile, as the case may be, and no handoff is necessary as long as the mobiles served by different beams using the same frequency do not cross each other.

Major limiting factor in mobile communications systems is co-channel interference. An adaptive antenna system has tremendous capability to suppress it using adaptive nulling process. Fading effect can be reduced using adaptive antenna system.

1.3 Smart antenna approaches

Three main categories of smart antennas may be defined based on how they produce their response: switched beam, direction finding, and optimum combining. The switched beam method employs a grid of beams and usually chooses the beam which gives the best SNR For direction-finding techniques (or spatial reference techniques, as they are also called) all the processing is focused on the acquisition and tracking of one parameter, the directions of the users. With optimum combining the output signal-to-interference-plus-noise ratio (SINR) is the parameter optimized. Table 1 summarizes the most important advantages and disadvantages of these techniques. When employing a smart antenna with one of the above mentioned methods in the general sense of a spatial filter, one further categorization can be recognized: spatial filtering for interference reduction (SFIR) and space division multiple accesses (SDMA). With SFIR the goal is, as with a traditional Omni directional or sectorised antenna, to support one user in each of the co-channel cells of the reuse pattern employed, and through interference reduction in the spatial domain to achieve a lower cell repeat pattern (reuse distance) With SDMA an adaptive antenna system is deployed in such a way that multiple users within the same cell can operate on the same time channel by exploiting the spatial separation of the users. This concept can be seen as a dynamic (as opposed to fixed) sectorisation approach in which each mobile defines its own sector as it moves. The major advantages and disadvantages of these techniques are summarized in Table.

	Advantages	Disadvantages
Switched beams	Easily deployed Tracking at beam switching rate	Low gain between beams Limited interference suppression False locking with shadowing, interference and wide angular spread
Direction finding	Tracking at angular change rate No reference signal required Easier downlink beamforming	Lower overall CIR gain Susceptible to signal inaccuracies, needs calibration Concept is not applicable to small cell NLOS environments
Optimum combining	Optimum SINR gain No need for accurate calibration Performs well even when the number of elements is smaller than the number of signals (MMSE approach)	Difficult downlink beamforming with FDD and fast TDD Needs good reference signal for optimum performance Requires high update rates

Table1:- Advantages and disadvantages of different smart antenna approaches

A smart antenna system relies heavily on the spatial characteristics of the operational environment to improve the output signal. For big cell structures (macro cells), there are three main scattering sources.

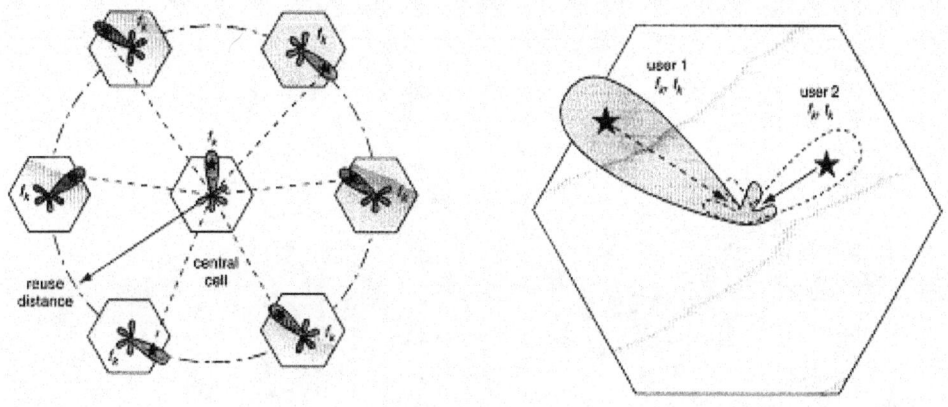

Fig 1:- (a) SFIR & (b)SDMA Concepts

Advantages and disadvantages of SDMA and SFIR:

	Advantages	Disadvantages
SDMA	No need for revised frequency planning to exploit capacity gain Single cell deployment for local capacity improvement	Requires discrimination between intracell SDMA users More complex radio resource management (angle and power)
SFIR	No need for major air interface changes Minor or no changes to the radio resource management	Relies on intelligent intracell handover Large deployments necessary to exploit the full capacity potentials

Table 2: Advantages and disadvantages of SDMA and SFIR

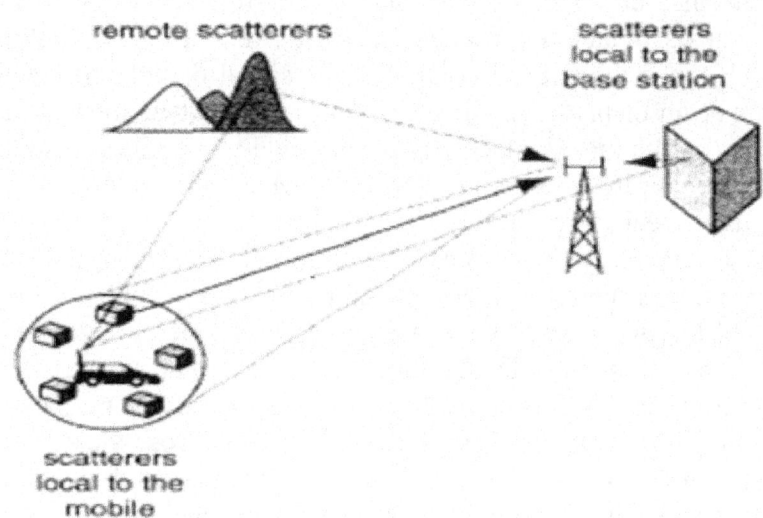

Fig 2:- Large Cell propagation environment

a. Scatterers local to the mobile: if the mobile is moving, these cause Doppler spread (i.e. time selectivity), and they also cause small delay and angle spreads.

b. Scatterers local to the base station: these contribute multipath rays with small delay spread and large angular spread.

c. Remote scatterers: these cause independent fading on paths and contribute multipath with large delay spread (frequency selective fading) and large angular spread.

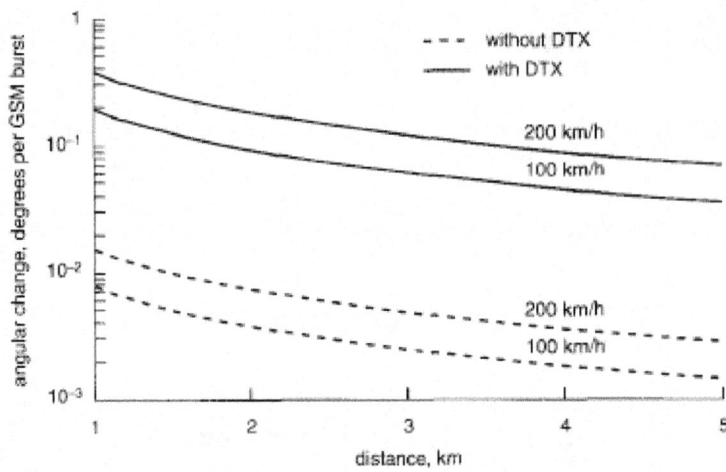

Fig 3:- Change in direction as a function of distance and speed for large cells, without DTX (1 burst /4.6 ms) and with maximum DTX (1 burst /120 ms).

An interesting point is that in big cell environments the angular change of the incoming signal depends generally on the velocity and the distance of the mobile from the base station (this is true when the multipath due to remote scatterers (Fig. 2) either doesn't exist or can be ignored by the adaptive algorithm). As an example, let us address the point concerning the angular change of incoming signals in a big cell environment in the context of a GSM system application. GSM is a time division multiple access radio standard with a slotted waveform structure having a 577 ns burst period and a 4.6 ms period between consecutive bursts. In this case, even with a worst case scenario of a mobile user traveling at 200 km/h (e.g. fast train) around a circle of radius 1 km, the angular velocity is less than 3.1 degree/ second, which corresponds for example to a 0.014 angular change between two consecutive GSM bursts, as shown in Fig. **3**. Since the angular change of the signal dictates the required update rate of the main beam then, obviously, the wider the beams are, the smaller the update rates required. Furthermore, since it is possible to reduce substantially the transmission rate either in the up or the down link when there is little or no speech (on average each user speaks less than half of the time during a normal conversation), many existing systems employ discontinuous transmission (DTX) in order to exploit this feature. Because discontinuous transmission effectively reduces the information rate (e.g., in GSM, from 1 burst every 4.6 ms, as mentioned above -no DTX to 1 burst every 120 ms - maximum DTX), this would mean that the angular change rate would increase considerably In small cells on the other hand, the propagation scenario is quite different. Here, there are many local scatterers in close vicinity to the mobile and the base station which result in much wider angular spread, low delay spread and moderate (pedestrian) to medium (mobile) Doppler spread. The previously used mapping through the spatial response (power versus angle) or 'spatial signature' of a big cell environment to a single 'dominant' direction cannot be done anymore because there is no connection between the direction of the signals and the physical location of the user, except from the direct paths in LOS cases". The 'spatial signature' itself has to be used for user location, but in this case there is no connection with the physical user location. Another very important issue for adaptive antennas is the downlink transmission. In time division duplex (TDD) systems, the up and down links can be considered reciprocal, provided that the channel characteristics have not changed considerably between the receive and transmit slots, i.e. there

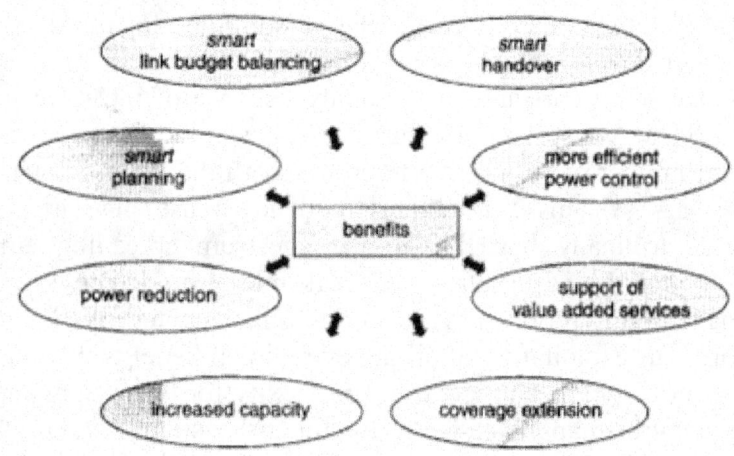

Fig 4:- Operational Benefits achieved with Smart Antenna.

is limited user movement between transmission and reception. Under this condition, the weights calculated by the adaptive antenna for the uplink can also be used for the downlink to achieve spatially selective filtering. Application of adaptive antennas in the downlink for frequency division duplex (FDD) systems seems to be one of the challenges related to this technology. The fundamental difference is that in FDD systems the downlink fading characteristics are independent of the uplink characteristics due to the frequency difference (typically -40 MHz), while the angles of arrival of the multipath rays remain the same. For current FDD systems such as GSM, DCS1800, IS54, IS95, the processing performed in the uplink can not be exploited directly in the downlink without any additional processing? Several approaches have been proposed in the literature which attempt to solve the downlink problem; some of them are:

a. Transmit diversity, where multiple base station antennas transmit delayed versions of the signal in order to create frequency selective fading at a single antenna at the receiver (mobile), which uses an equalizer or RAKE receiver to obtain diversity against fading.

b. Transmission with feedback from the mobile. Here feedback from the mobile to the base station is used to estimate the impulse response of the radio channel, hence traditional diversity schemes or beam forming can be employed.

c. estimation of the transmit spatial correlation matrix from parameterization of the receive spatial correlation matrix. One such parameterization can be performed for the angles of arrival, which can subsequently be calculated by direction-finding algorithms.

d. subspace processing (a similar idea to c). Whereas *a.* and *b.* rely on the separation of the instantaneous channel vectors, with this method separation of the channel subspaces (second order spatial correlation matrix) is employed. The reason for such processing is that a subspace is a much more stable entity than a channel vector and hence, instead of tracking the instantaneous channel, subspace beam forming methods track the subspace structure. Other methods which exploit particular characteristics of the air interface method employed can also be used to compensate for the possible imbalance caused by the downlink problem. For example, one such method could be adaptive resource allocation. This effectively means allocating more radio channels (TDMA) or bandwidth (CDMA) for the downlink so that the benefits are balanced between the two links. Furthermore, techniques which convert one form of diversity at the base station to another form at the mobile - space/time, space/path, space/coded modulation, etc. This can also offer alternative solutions for the downlink problem. Different air interface techniques have different impacts on the design and the optimum approach for smart antennas, mainly because of the different interference scenarios. With TDMA systems, a frequency reuse pattern is usually employed, which leads to a small number of strong interferers for both the up and down links (usually 2-4). With CDMA systems a total (1) frequency reuse plan is assumed, which effectively leads to many but, due to the spectrum spreading employed, weak interferers in the uplink and usually 6-12 weak interferers in the downlink. Possibly the most challenging problem related to smart antennas is their practical implementation. Full exploitation of all the operational benefits that will be described in the next section would mean increased complexity for the system and hence would require a fully integrated approach in an 'intelligent' system. RF and DSP technology will have to evolve further before this can be achieved in a cost-effective manner. Nevertheless, two facts indicate that this is still a promising and valid way forward: first, the fact that there are partial or efficient implementations that can be employed in real systems to reduce complexity and, second, the fact that by the time mobile communication systems are ready to fully support smart antennas (most probably with 3rd generation systems) the required technology will be available and mature enough to fully support adaptive antennas (assuming Moore's law). Furthermore, although current implementation and installation costs are thought to be high for smart antenna systems in comparison with traditional Omni directional or sectorised schemes, if the range-capacity gains and the other benefits that will be discussed in the next section are included in the overall cost calculations, then it becomes obvious that this technology is rather economical even for today's systems.

1.4 Operational benefits with smart antennas

In the previous section Fig. 4 shows an overview of the operational benefits that can be achieved from the deployment of an adaptive antenna in a mobile communication network'. Each of these benefits is discussed in greater detail below. The diversity gain offered by an antenna array reduces the fading of the radio signal and hence the power control requirements are eased. With information about the location and speed of a mobile user, the decision is as to which cell to hand the user to be easier task. Combination of this kind of information with the handover process can ultimately lead to a 'smart' instead of 'soft' or 'hard' handover. Furthermore, in order to support the soft handoff quality enhancement technique for DSCDMA within a mixed cell environment, all cell types have to operate on identical carrier frequencies. One possible way of achieving the RF power balancing within each area of a mixed cell scenario needed to provide seamless handover and simultaneously avoiding the near-far effect is to exploit the spatial filtering properties offered by an adaptive antenna.

1.5 Support of value added services (Better signal quality - higher data rates)

In noise or interference limited environments, the gain that can be achieved with an antenna array can be exchanged for signal quality enhancement, i.e. lower BER. This is demonstrated in Figs. 5 and 6. Fig. 5 shows the probability of detection for the case where a matched filter (a detector which maximizes the probability of detection with

9

additive noise) is employed for each array element. The probability of detection in this case is:

$$P_D=Q(Q^{-1}(P_F)-\sqrt{(M*SNR)})\ldots\ldots\qquad(1)$$

Where $Q(.)$ is the Q function, P_D is the probability of detection, P_F is the threshold probability of false alarm, M is the number of elements and SNR the signal-to-noise ratio. It can be seen that the detection performance can be significantly enhanced with an antenna array. For example, for 0 dB SNR, there is 10% probability of detection with a single element, 20% with two elements, 35% with four, 65% with eight, 85% with twelve, 95% with sixteen and almost 100% with twenty. Fig. 6 shows another example, this time for a DSCDMA system, where the following approximate formula for the BER is used:

$$P_B=Q\sqrt{(3*SF*SIR_{omni}*K^{-1})}\ldots\qquad(2)$$

Where SF is the spreading factor, SIR_{omni} is the signal-to-interference ratio with an Omni directional antenna, and with BW and SLL the ideal or effective beam width (degrees) and average side lobe level (linear values).

$$K=BW/360+SLL(1-BW/360)\ldots\ldots\ldots(3)$$

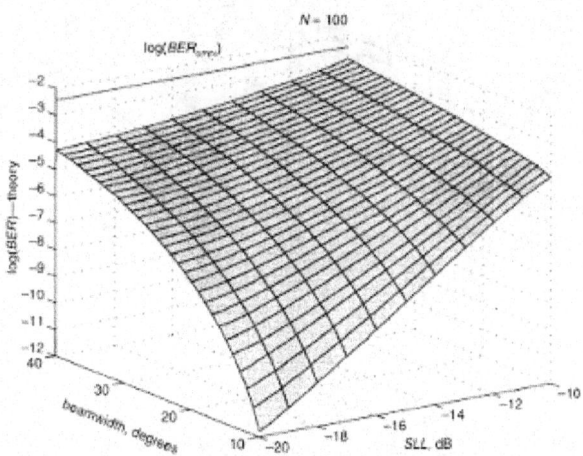

Fig 5:- BER with an adaptive antenna as a function of achieved beamwidth and average side-lobe level.

Fig. 5 shows a plot of equation: 2 for the case of 100 users per cell (N) in a multi tier system (4 tiers) for values of achieved beam width and side lobe level between 10° and 30°, and -10 dB and -20 dB, respectively. As an example from Fig. 6, if the smart antenna that is employed at the base station of the central cell can achieve a radiation pattern with a beam width of 20° (ideal or effective), then an improvement of 1-7 orders of magnitude for the BER can be accomplished with average side lobe levels (ideal or effective) between-10 dB and -20 dB, respectively. Obviously, the side lobe level achieved has such a profound effect on the BER results because it reduces the interference received at that region and hence improves the overall SIR, as seen from equations: 2 and 3. Furthermore, through spatial filtering of the multipath at the base station (BS) and/or the mobile station (MS), the RMS delay spread of the channel can be reduced, which is a positive thing if the objective is to achieve higher data rates. This advantage increases in environments where the angular spread of the multipath is wide, i.e. in small cell scenarios. Examples of RMS delay spread reduction with smart antennas in outdoor micro- and indoor Pico cells are shown in Fig. 6. The method used to produce the results shown in Fig. 6 comprised using the ray tracing tools which produce the impulse response of the radio channel, then steering the antenna pattern (15 dBd, 26°3 dB beam width) from 0° to 360° in steps of 5° and for each step calculating the power and the RMS delay spread. The results are for an LOS (line-of sight) point. The relationship between high power and small RMS delay spread can be seen in these figures (e.g. 270°-300° for Fig. 7a and 180°-210° for Fig. 7b), although it must be mentioned that it is not always straightforward, especially for NLOS situations. In terms of RMS delay spread, the directional antenna has an advantage, both in LOS and NLOS (non line of sight) situations, when it is steered towards the direction of high power. When the main beam of the directional pattern is miss pointed towards other than the peak power directions, more multipath rays are received from the base station, and this increases the RMS delay spread.

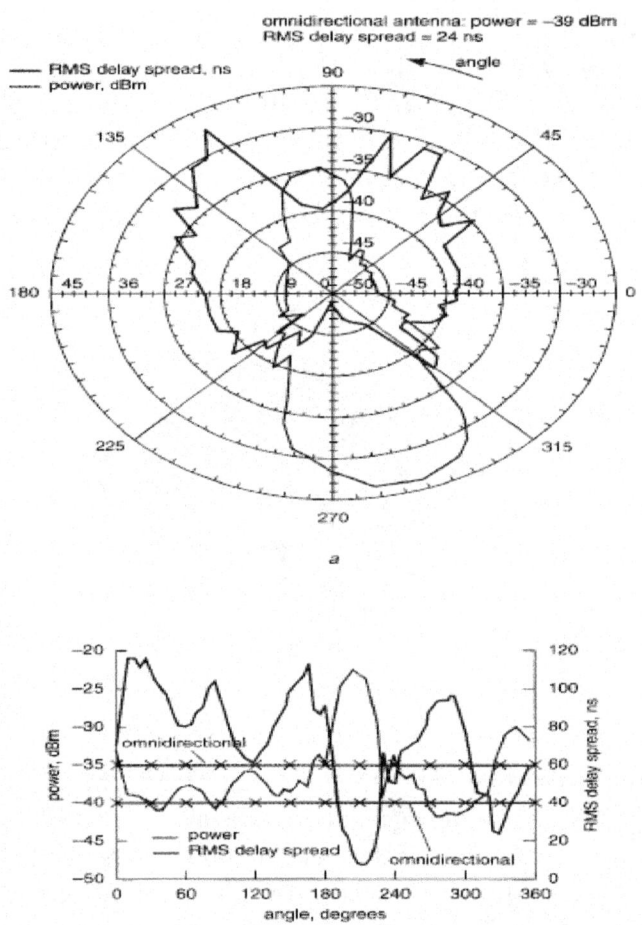

Fig 6:- RMS delay spread reduction with adaptive antennas in *(a)* microcells and *(b)* picocells

Ability to support user location for emergency calls

Adaptive antennas can provide user location information (direction-finding methods are naturally the most suitable for this application).

Location of fraud perpetrators

A significant portion of the wireless operator's revenues is lost every year due to fraud. Many technologies currently applied in order to solve this problem are effective in identifying the fact that fraud is occurring but none of them provides the operator with a long-term solution, i.e. the ability to remove the criminal from the activity, instead of deactivating the phone number from the switch. Since adaptive antennas can provide user location information, this now becomes feasible. *Location-sensitive billing* Adding a third dimension - location - to the current two dimensions for charging rates (usage against time of day, i.e. peak, off-peak) will provide an operator with the ability to control its network by encouraging (or discouraging) any type of usage behavior, and ultimately

Vehicle and fleet management

Current vehicle management techniques usually suffer from the cost disadvantage of having a single-purpose transmitter network. Location and subsequent navigation of a vehicle is something that can be done with an adaptive antenna. Possible extensions to this idea could also be package monitoring and stolen vehicle recovery.

Optimum smart system planning

Today the planning of a network is typically theoretical and once implemented is verified through limited testing. One important element missing from this planning process is that designers never really know where wireless devices are located when a specific cell site is handling a problematic call. Obviously this situation can be greatly improved with the user location capabilities of adaptive antennas. Furthermore, with an adaptive antenna an advanced intelligent network, which will lessen the planning burden and improve the network efficiency based on a number of criteria optimized through the intelligence provided by the base station antenna, becomes possible. Optimization parameters include intra- and intercellular reuse planning, link balancing requirements, handover requirements, traffic density and many others. When such intelligence at the base station is enhanced by intelligence at the network level, self-configuration and overall optimization becomes viable.

Coverage extension

Adaptive antennas can increase the network coverage through antenna directivity and interference reduction. The gain G (assuming an antenna efficiency of 100% and no mutual coupling) that can be achieved with an antenna array of M elements is (it can also be proved that this represents the SNR improvement) :
$$G \sim 10 \log_{10} M \ldots\ldots (4)$$
The additional gain (compared to the standard element gain) can obviously be exploited for range extension. With small angular spread and single slope path loss with exponent n, the range extension factor, **REF,** can be calculated as:
$$REF = r_2/r_1 = M^{1/n} \ldots\ldots\ldots\ldots\ldots (5)$$

Or, with the Hata path loss model:
$$REF = M^{.3} \ldots\ldots (6)$$
Where **r,** and **r,** are the default (with a single element) and extended (with multiple elements) ranges, respectively. The area improvement factor, **AEF,** is:
$$AEF = (r_2/r_1)^2 = REF^2 \ldots\ldots\ldots\ldots\ldots\ldots (7)$$
The inverse of the area extension factor represents the reduction factor in the number of base stations *(BSRF)* needed to serve the same area as with a single element. Obviously, one can get the full range extension from the directivity of the array when **the** angular spread of the signal is less than the antenna beam width (otherwise the desired signal is reduced). Fig. 7 shows that with ten antenna elements (10 dB of SNR improvement) the range can be almost doubled and the area almost quadrupled, or the number of base stations can be reduced to almost 25% of the original number. Also, it

can be seen that with smaller path loss exponents (the curve for **n=3** in Fig. 7) the improvement factors will be higher, i.e. the improvements will be higher in LOS as compared to NLOS scenarios. Also, the calculated range extension with the Hata path loss model is between the predictions for path loss exponents of 3 and 4.

Reduced transmit power

Limitations on the maximum EIRP (effective isotropic radiated power) introduced by standards could mean that the array gain cannot be exploited for coverage or area extension. Also, recent public worries over health issues originating from exposure to electromagnetic radiation (no matter how reasonable or unreasonable this may be) will almost certainly force the governing/standardization bodies to change the current radiation standards in the future and adopt a lower emission policy, in particular for the mobile handsets. In such cases, one could exploit the base station array gain to reduce the power transmitted by the mobile. This reduction is also beneficial because it relaxes the battery requirements and hence it can increase the talk times or reduce the size/weight of the handsets. Furthermore, if the received power requirement at the mobile remains the same with an M element array at the base station, then the output power from the base station power amplifiers (PAS) can be reduced by M^2, which will reduce the total transmitted power from the array by M^1. The explanation for this is that the base station now employs MPAs and can simultaneously exploit the directivity of the array. Ten antenna elements will reduce the total transmitted power by 10 dB, while the output of each PA will be reduced by 20 dB. The latter has obvious cost implications, since the high-power amplifiers are expensive hardware components of a system.

Smart link budget balancing

Examination of the up/down link budgets reveals an imbalance which originates from the difference in the power amplifiers employed at the two ends. This imbalance is usually around 10-20 dB and its actual value depends on the PA used at the base station site. As an example consider a case where there is a 13 dBW PA with a 17 dBi antenna, giving 1 kW EIRP at the base station (this example ignores losses from cables, connectors, mismatching, etc. and fading margins at the mobile and the base station). The mobile might be able to achieve 1 W EIRP, which, together with the 17 dBi from the base station antenna gain, will produce a deficit of more than 13 dB and, hence, a link imbalance. Given that the antenna gains can be exploited at both ends, it becomes obvious that the up and down links can be balanced by using the additional directivity offered by an adaptive antenna at the base station site. For the case of a power amplifier with a dBW output, ξ dB gain from high sensitivity receivers and other improvements (e.g. employing diversity at the mobile), and $EIRP_{MS}$ dBW at the mobile, the gain from the adaptive antenna at the base station in order to balance the two links should ideally be capabilities of a smart antenna this concept can be extended and 'smart' link management can lead to dynamically optimized power budgets for different scenarios.

$G(dB) = a - \xi - EIRP_{MS}$. By exploiting the adaptive

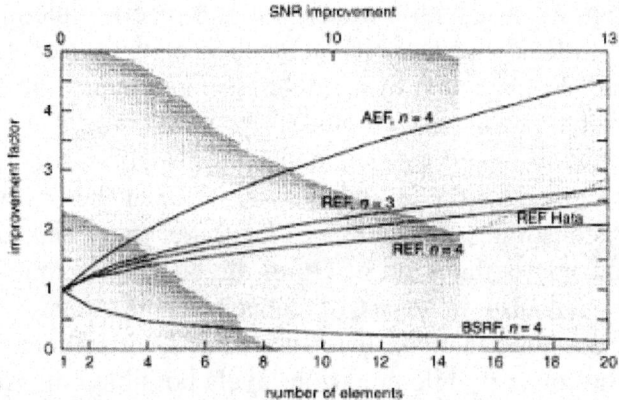

Fig 7:- Range, area and base station reduction factors as a function of the number of elements and SNR improvement

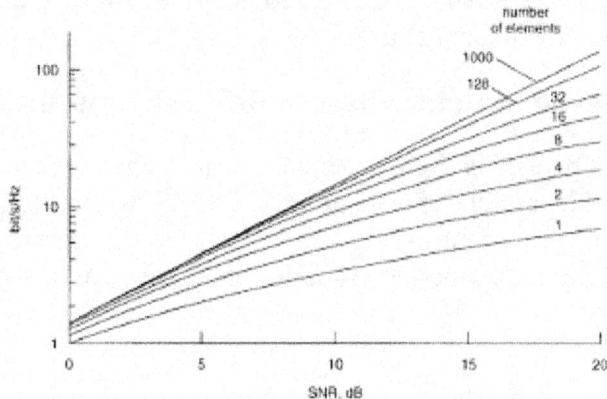

Fig 8:- Spectrum efficiency limit as a function of the SNR and the number of antenna elements

Increased capacity

Adaptive antennas can provide capacity increases through several mechanisms. Starting with Shannon's expression for the capacity of a channel with bandwidth Wand with additive Gaussian noise,

$$C = W\log_2(1+SNR) \text{ bit/s} \ldots\ldots\ldots\ldots\ldots\ldots\ldots\ldots\ldots \quad (8)$$

It can be shown35 that with an antenna array which divides the power equally into *M* parts and sends it into M parallel independent channels *(M is the number of antenna elements)*, the capacity in one channel is now:

$$C_{array} = \sum_{k=1}^{M}[W\log_2(1+SNR/M)] = M*W\log_2(1+SNR/M)bit/\sec \ldots\ldots\ldots(9)$$

Fig. 8 shows the spectrum efficiency (bit/s/Hz) as a function of the SNR and the number of antenna elements. Apart from the increase of the spectrum efficiency as the number of array elements increase, it can be seen that as the number of elements becomes very large, the capacity of the channel becomes simply proportional to *SNR*. Extending this concept by multiplexing *M* transmit and N receive channels can offer significant gains. It is shown that the theoretical capacity is now (flat fading, stationary propagation environment) :

$$C_{M,N} = W\log_2 \det[I_N + SNR/M \ HH^T] \quad \text{bits/s} \ldots\ldots\ldots\ldots(10)$$

Where 'det' means determinant, I_N is the $N*N$ identity matrix and $\mathbf{H^T}$ is the conjugate transpose of the normalized (spatial average power loss normalized to unity) channel matrix (the i j th element of which is the transfer function of the j th transmitter to the I th receiver). It is shown that with 8 transmit and 8 receive antennas and 1% outage probability with 21 dB averages SNR at each receiving element, the maximum capacity is more than 40 times that of a single antenna element at the transmitter and the receiver. Furthermore, a layered space /time architecture can achieve such capacities. With an Omni directional antenna only a small portion of the transmitted power is actually received by the intended user, while at the same time most of the transmitted power constitutes interference for other potential users. Hence, the problem is dual: not only is this kind of Omni directional communication inefficient in terms of power, but also in terms of capacity. One practical way to increase capacity is to decrease the radiated power associated with directive transmission in combination with the lower mobile emission levels possible with directive reception. In other words, by exploiting the spatial filtering that an antenna array offers, it is possible with the help of an adaptive method to confine the radio energy associated with a given mobile to a small addressed volume, thus reducing interference experienced from and to co-channel users.

1.6 Brief Overview of Satellite based Mobile Communication

"Mobile" or "mobile phone" is used to denote a communications device on the move, including a hand-held portable and a mobile vehicle on land (also known as land mobile), a ship, or an aircraft. To begin, let us see how a typical mobile communications system involving land mobiles and base stations works

Call Initiation from a Mobile

When a mobile phone is switched on, it scans the control channels and is tuned to the channel with the strongest signal, usually arriving from the nearest base station. The phone user identifies itself and establishes authorization to use the network. The base station then sends this message to the switching center connected to the telephone network, which controls many base stations. It assigns a radio traffic channel to the phone under consideration, as the control channels are used by all phones in that area and cannot be used for data traffic.

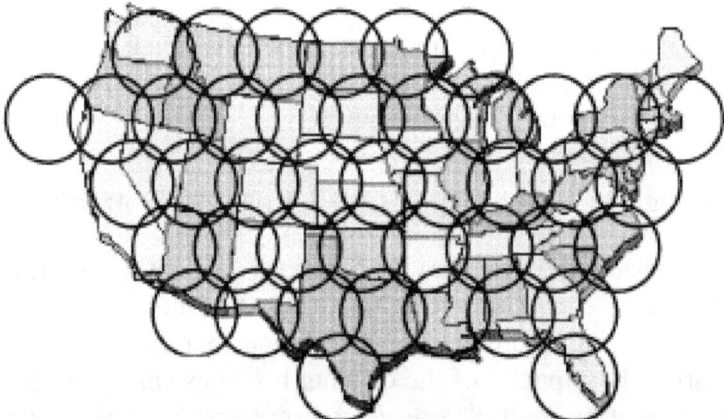

Fig 9:- A typical setup of a satellite mobile coverage

Once the traffic channel is assigned, this information is relayed to the phone via the base and the phone tunes itself to this channel. The switching center then completes the rest of the call.

15

Initiation of a Call to a Mobile

When someone calls a mobile phone, the switching center sends a paging message through several base stations. A phone tuned to a control channel detects its number and responds by sending a response signal to the nearby base, which then informs the switching center about the location of the phone. The switching center assigns a channel, and the call is completed.

Multiple-Access Schemes

The range of frequencies available for mobile communications is utilized in a number of ways, referred to as multiple-access schemes. Three basic schemes are FDMA, TDMA, and CDMA. The standard analog FDMA scheme allocates different carrier frequencies to different users. A TDMA scheme, useful for digital signals, allocates different time slots to different subscribers using the same carrier frequency and thus interleaves signals from various users in an organized manner. Traffic in the base-to-mobile direction is separated from that in the mobile-to-base direction by either using different carrier frequencies or alternating in time. The two schemes are referred to as FDD and TDD. The use of separate frequencies for transmission in both directions does not require as precise a synchronization of data flowing in the two directions as does the alternative transmission method. Details on various aspects of the TDMA scheme, including the capacity of a system using this scheme, may be found in the various literature. A CDMA scheme, on the other hand, is a spread spectrum method that uses a separate code for each user. These codes are large PN sequences that spread the spectrum over a larger bandwidth, simultaneously reducing the spectral density of the signal. Various CDMA signals occupy the same bandwidth and appear as random noise to each other. In theory, the capacity provided by the three multiple access schemes is the same and is not altered by dividing the spectrum into frequencies, time slots, or codes, as explained in the following example. Assume that there are six carrier frequencies available for transmission covering the allocated spectrum. In a system using the FDMA scheme, six frequencies are assigned to six users, and six simultaneous calls may be made. TDMA generally requires a larger bandwidth than FDMA, so a system using this scheme creates two TDMA channels and divides each into three time slots, serving six users. A CDMA channel requires a larger bandwidth than the other two and serves six calls by using six codes, as illustrated in Fig. 10. In practice, however, the performance of each system differs, leading to different system capacities. Furthermore, each scheme has its advantages and disadvantages, such as complexities of equipment design, robustness to system parameter variation, and so on. For example, TDMA processes signals from all users simultaneously, requiring complex time synchronization of the different user data. This is not the case for CDMA, which processes individual data independently at the receiver. It does, however, require code synchronization. It is argued that although there does not appear to be a single scheme that is the best for all situations, the CDMA scheme possesses characteristics that give it distinct advantages over other schemes. The SDMA scheme, also referred to as space diversity, uses an array of antennas to provide control of space by providing virtual channels in an angle domain. Using this scheme, simultaneous calls in different cells can be established at the same carrier frequency.

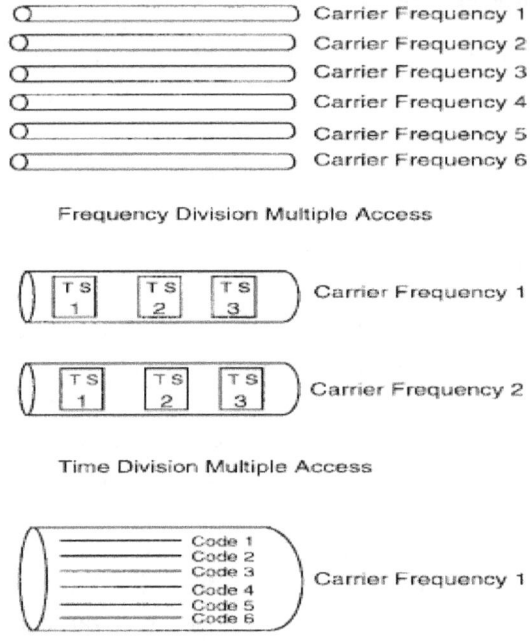

Fig 10:- Channel usage by FDMA TDMA & CDMA

The SDMA scheme is based upon the fact that a signal arriving from a distant source reaches different antennas in an array at different times due to their spatial distribution, and this delay is utilized to differentiate one or more users in one area from those in another area. The scheme allows an effective transmission to take place in one cell without disturbing the transmission in another cell. Without the use of an array, this is accomplished by having a separate base station for each cell and keeping cell size fixed, whereas using space diversity, the shape of a cell may be changed dynamically to reflect the user movement. Thus, an array of antennas creates an extra dimension in this arrangement by providing dynamic control in space

Propagation Characteristics

An understanding of propagation conditions and channel characteristics is important to the efficient use of a transmission medium. Lately, attention has been given to understanding the propagation conditions where a mobile is to operate, and many experiments have been conducted to model the channel characteristics.

Fading Channels

The signal arriving at a receiver is a combination of many components arriving from various directions as a result of multipath propagation. It depends upon terrain conditions and local buildings and structures, causing the received signal power to fluctuate randomly as a function of distance. Fluctuations on the order of 20 dB are common within the distance of one wavelength. This phenomenon is called fading. One may think of this signal as a product of two variables. One of the components referred to as the short-term fading component, changes faster than the other one and has a Rayleigh distribution. The second component is a long-term or slow varying quantity and has log- normal distribution. In other words, the local mean varies slowly with log-normal distribution, and the fast variation around the local mean has Rayleigh distribution. The fluctuation in the local mean is caused by shadowing and

17

thus is referred to as shadow fading, whereas the fast change in signal amplitude is caused by the phase differences in signal components and is referred to as multipath fading. A stationary subscriber may also observe fading when the differential phases of various multipath components change fast with frequency. Such fading is referred to as frequency-selective fading. When the fading is independent of frequency, it is referred to as flat fading. A movement in a mobile receiver causes it to encounter fluctuations in the received power level. The rate at which this happens is referred to in mobile communications literature as the fading rate, and it depends upon the frequency of transmission and the speed of the mobile. For example, a mobile on foot operating at 900 MHz would cause a fading rate of about 4.5 Hz, whereas a typical vehicle mobile would produce a fading rate of about 70 Hz.

Doppler Spread

The movement in a mobile also causes the received frequency to differ from the transmitted frequency due to the Doppler shift resulting from its relative motion. As the received signals arrive along many paths, the relative velocity of the mobile with respect to various components of the signal differs, causing the different components to yield different Doppler shifts. This can be viewed as spreading of the transmitted frequency and is referred to as the Doppler spread. The width of the Doppler spread in frequency domain is closely related to the rate of fluctuations in the observed signal.

Delay Spread

Due to the multipath nature of propagation in the area where a mobile is being used, it receives multiple and delayed copies of the same transmission, resulting in spreading of the signal in time. The delay spread may range from a fraction of a microsecond in urban areas to something on the order of 100 s in a hilly area, which restricts the maximum signal bandwidth between 40 and 250 kHz. This bandwidth is known as coherence bandwidth. The coherence bandwidth is defined as the inverse of the delay spread. For digital modulated schemes, the signal bandwidth is the inverse of the symbol duration.

For a signal bandwidth above the coherence bandwidth, the different frequency components of the signal arrive at a receiver at different times, and the channel becomes Frequency selective. Frequency-selective channels are also known as dispersive channels, whereas the non-dispersive channels are referred to as flat fading channels. A channel becomes frequency selective when the delay spread is larger than the symbol duration and causes inter-symbol interference (ISI) in digital communications. The ISI may be reduced to a certain degree by using equalizers in TDMA and FDMA systems.

Link Budget and Path Loss

Link budget is a name given to the process of estimating the power at the receiver site for a microwave link, taking into account the attenuation caused by the distance between the transmitter and the receiver. This reduction is referred to as the path loss. In free space, the path loss is proportional to the second power of the distance, that is, the distance power gradient is two. In other words, by doubling the distance between the transmitter and the receiver, the received power is reduced to one-fourth of the original amount. For a mobile communications environment utilizing fading channels, the distance power gradient varies and depends upon the propagation conditions. Experimental results show that it ranges from a value lower than two in indoor areas with large corridors to as high as six in metal buildings. For urban areas, the path loss between the base and the cell site is often taken to vary as the fourth power of the distance between the two. Normal calculation of the link budget is made by calculating the C/N ratio where noise consists of background and thermal noise and the system utility is limited by the amount of this noise. However, in mobile communications systems, the interference due to other mobile units is a dominant noise compared to the background and man-made noise. For this reason, these systems are limited by the amount of total interference present rather than the background noise, as in the other cases. In other words, the SIR is the limiting factor for a mobile communications system rather than the SNR, as is the case for other communications systems. The calculation of link budget for such interference-limited systems involves calculating the carrier level above the interference level contributed by all co-channel sources.

Channel Assignment

The generic term "channel" is normally used to denote a frequency in the FDMA system, a time slot in the TDMA system, and a code in the CDMA system, or a combination of these in a mixed system. Two channels are different if they use different combinations of these at the same place. For example, two channels in an FDMA system use two different frequencies. Channel assignment is a complex process where a finite number of channels are assigned to various base stations and mobile phones. In a system with fixed channel assignment, channels are assigned to different cells during the planning stage, and the assignment rarely changes to reflect the traffic needs. A channel is assigned to a mobile at the initiation of the call and the mobile communicates with the base using this channel until it remains in the cell. Dynamic channel assignment, on the other hand, is an efficient way of channel usage in a multiple-user environment. With this arrangement, a channel with the minimum interference is found before assignment. The interference level of all the channels used and unused is monitored periodically, and the channel assignment during the call may be changed from the one with high interference to the one with low interference, the so-called quiet channel. As the interference environment is constantly changing due to the movement of mobiles, this ensures that the performance of the system is not affected adversely as long as there are quieter channels available. Monitoring traffic and searching for quieter channels is a complex process with heavy computational demand. A study of a dynamic channel-assignment scheme based upon neural networks shows the possibility of reducing this complexity without degrading the handoff performance compared to conventional methods and argues about the suitability of neural networks for dynamic channel assignment. Though it is not necessary that the neural networks always provide the optimal solution to the assignment problem, some improvement in channel utilization is achievable, as reported in various literatures. There exist a number of other channel assignments between the two extremes of fixed and dynamic, including flexible channel assignment and channel borrowing schemes. Both schemes are a variation of the fixed channel assignment. In the former, the assignment is periodically altered to reflect traffic needs, whereas in the latter, unused channels in a cell are borrowed by a congested cell. The total transmitted power of the base station sometimes is also considered as a criterion for channel allocation.

Handoff

A mobile phone's movement may cause it to be far away from the original base station through which the phone is connected to the switching center. This distance is detected by the switching center monitoring the signal strength arriving from the phone at the base station. Once the signal becomes too weak, the switching center reassigns a new traffic channel via a base station closer to the phone and asks the phone to tune to this new channel. This is known as handoff or handover, a process that is generally transparent to the mobile user. In a typical call, there may be several handoffs. The handoff needs to be performed based upon certain established policy, which may involve the measured power level and the quality of reception. The mean signal strength decreases as a mobile moves away from a base, whereas it increases when a mobile approaches a base. A handoff policy based upon this fact may use the relative strength of signals received from various base stations. The measured as well as predicted signal strength may be involved in the decision making. A policy based upon the quality of reception may involve call blocking probability as well as handoff blocking probability. A policy based upon a minimum number of handoffs may lead to poor communication quality. A good handoff scheme should aim to minimize the effects of handoff on calls, including the noticeable disturbances that may be severe in some services, such as live video transmission. In a system where different cells use different frequencies, the handoff requires switching frequencies and is referred to as a hard handoff. In systems where all cells use the same frequency but different codes, as in the case of CDMA, only codes need to be switched, and the process is known as a soft handoff. Soft handoff can be used between sectors of the same base station. It is also being proposed for satellite-based mobile communications systems.

Power Control

It is important that a radio receiver receive a power level that is high enough for its proper function but not so high as to disturb other receivers. To this end, two methods are used. One is concerned with maintaining a constant power level at the receiver by transmitter power control. In the other, the SNR is kept constant. In both the cases, the receiver controls the power of the transmitter at the other end. For example, a base would control the power transmitted by mobile phones, and vice versa.

Item	Macro-cell	Micro-cell
Cell radius	1-20 Km	Less than 1 Km
Transmitter power	1-10 W	Less than 1W
Channel Fading	Raleigh	Ricean
RMS delay spread	.1-10 micro sec.	10-100 ns
Max bit rate per channel	.3 Mbps	1Mbps

Table 3: Comparison of Micro cell and Macro cell System Parameters

A receiver monitors its received power—or the SNR, as the case may be—and sends the control signal to the transmitter to control its power as required. Sometimes a separate pilot signal is used for this purpose. Power control reduces the near-far problem in CDMA systems and helps to minimize interference near the cell boundaries when used in forward link.

Satellite Mobile Communications

A variety of configurations involving satellites for wireless communications exists. For example, in global positioning systems, signals received from a number of orbiting satellites are used to determine the position of a receiver. A direct broadcasting system uses satellites to transmit signals to television users. In both of these systems, the transmission is generally one way in nature. For a communications satellite system, on the other hand, a two way communication is required. Though the requirements for different systems—such as maritime, aeronautical, and land-mobile systems—may be different, the basic principle of operation remains the same. Basically, a satellite acts as a relay station between a mobile and the base station, also known as a hub station. A geostationary satellite may provide a large coverage similar to the large and fixed coverage of a mobile radio communications system, or it may provide a number of spot beams with different beams using different frequencies, as in cellular radio. The communications satellites tend to complement the terrestrial network and play a major role in areas where the latter is not competitive or is underdeveloped. Goes' have a number of drawbacks when it comes to global voice communication. These satellites are placed around 36 000 km above the earth, and two-way propagation delay on the order of 0.6 s encountered by a signal makes it unacceptable for voice communication. There also exists a difficulty in covering the area of the globe far south and far north of the equator using Goes'. There are problems in these regions as the satellite appears close to the horizon. Even using large antennas, communication beyond 75 north and south of the equator is not satisfactory. Blockage of the satellite by large buildings, even when one is situated at about 40–45 north or south of the equator, is also a problem. The communications system requires high-power transmitters with large antennas to overcome the propagation loss suffered by the signal due to a large distance. Such conditions are not practical for mobile systems. To overcome these problems, a number of low-orbit satellites have been proposed for mobile satellite communications systems. These include LEOS's at an altitude of around 1000 km, MEOS's at an altitude of around 10 000 km, and HEOS's with varying altitudes. Though LEOS and MEOS require less power and cause less delay, making portable mobiles viable, a large number of satellites are required for global coverage. There is a requirement of fast handoffs, as the satellites move rapidly and are in view only for a short time. There is also a concern of limited lifetime (five–ten years) due to orbital decay, requiring regular replacement. A number of satellite communications systems are planned and will be in operation in the near future.

Among these are the ARIES, GLOBESTAR, IRIDIUM, and ODYSSEY systems, which will provide full global coverage with voice, data, and fax services. These are known as big-LEO systems. The word "big" indicates that these satellites would have enough power and bandwidth to provide near toll-quality voice services to hand-held portable and vehicle mobiles, along with other services. Little-LEO systems, such as LEOSAT, STARNET, and VITASET, would provide global coverage with non-voice services. These would provide low-bit-rate services and are expected to be of small size and low mass. The IRIDIUM system is briefly discussed here as a typical system to provide a feel for its operation, service provided, and other parameters. IRIDIUM is proposed by Motorola to provide global coverage for voice, data, fax, paging, and so on. It would have 66 satellites (LEO 780 km) in six polar orbit planes at 86.4 inclinations, each with a mass of 700 kg, and would be operational in 1998. Each satellite would provide 48 beams. This gives in total 3168 cells.

system	No of	Multiple	orientation	Altitude

	satellites	access		(KM)
ARIES	48	CDMA	CIRCULAR	1020
ELLIPSO	15	CDMA	ELLIPTICAL/CIRCULAR	7800
GLOBESTAR	48	CDMA	CIRCULAR	1400
IRIDUM	66	FDMA/TDMA	CIRCULAR	780
ODESSEY	12	CDMA	CIRCULAR	10350
TELEDESIC	840	FDMA/TDMA	CIRCULAR	700
ORBCOMM	26	FDMA	CIRCULAR	970
STARSYS	24	CDMA	CIRCULAR	1000
VITASET	2	FDMA	CIRCULAR	800

Table 4:- Comparison of parameters for some Mobile Satellite System

These would move over the surface of the earth as satellites move, and 2150 cells would be active at any time to cover the whole earth. The beams would be generated by three phased array panels, each generating 16 simultaneous shaped beams. The system would complement terrestrial systems. It would use 1616–1626.5 MHz for both uplink and downlink and 23 GHz for inter-satellite links. A mobile unit would first use the local cellular ground-base system and then a satellite using one of the frequencies. A satellite would demodulate the signal to get the address of the destination and then send the message to that satellite, which is in the viewing area of the destination cell. It would use FDMA for uplink and time division multiplexing for the downlink.

Progress and Trends in Mobile Communications

Mobile communications technology has come a long way since the pioneering work at AT&T Bell Laboratories during the 1960's and 1970's. There, researchers coupled the idea of frequency reuse with digital switching, leading to the opening of the first operational cellular system in Chicago in October 1983. The first-generation analog cordless phones and cellular systems became popular using the design based upon a standard known as AMPS. Similar standards based upon FDMA were developed around the world, including TACS, NMT 450, and NMT 900 in Europe; ETACS in the United Kingdom; C-450 in Germany; and NTT, JTACS, and NTACS in Japan. The handoff decision for these systems is based upon the power received at the base or, in the case of C-450, the round-trip delay. AMPS systems used in Australia, the United States, Canada, and Central and South America, use an 824–849 MHz band for transmission from mobiles to base and an 869–894 MHz band for transmission from base to mobiles. There are 832 channels of 30-kHz width. The system typically makes use of 12 frequency reuse plans with Omni directional antennas or seven frequency reuse plans with three sectors in each cell. In contrast to the first-generation analog systems, second-generation systems are designed to use digital transmission, to have a separate dedicated channel for exchange of control information between base and mobiles, and to employ TDMA or CDMA as a multiple-access scheme. These systems include the Pan-European GSM and DCS 1800 systems, the North American dual-mode cellular IS-54 system, the North American IS-95 system, and the Japanese PDC system. The GSM, DCS1800, IS-54, and PDC systems use TDMA, whereas IS-95 uses CDMA. Both the IS-54 and IS-95 systems are designed to operate in the frequency band used by the AMPS system. The first-generation analog cordless phones are designed to communicate with a single base, effectively replacing a telephone cord with a wireless link to provide terminal mobility in a small coverage area, such as within a house or office. The second-generation digital cordless systems are being developed with the aim of using the same terminal in residential as well as public-access areas, such as offices, shopping centers, and so on, and of being able to receive and originate calls. Digital cordless systems include CT2, a British standard originally adopted in 1987 and later augmented in 1989, the DECT standard, and PHS of Japan. CT2 employs an FDMA scheme, providing one channel per carrier, whereas the DECT and PHS systems are designed to use TDMA, providing 12 and four channels per carrier, respectively. The third-generation mobile communications systems are being studied worldwide under the names of UMTS and FPLMTS. The aim of these systems is to provide users advanced communications services having wideband capabilities and using a single standard. In third-generation communications systems, satellites are going to play a major role in providing global coverage. Adaptive antenna array processing has the potential to provide designers with an extra dimension of SDMA along with FDMA, TDMA, and CDMA in solving this mammoth task. Details on the use of arrays for mobile communications are provided in the remainder of this paper.

1.7 Satellites-Mobile System

In this system, the mobiles (in this case mostly vehicles) directly communicate with the satellite, in contrast to a system where a base station acts as a repeater station by communicating with the satellite on one hand and the mobiles on the other. In this case, multiple antenna elements may be utilized on a mobile as well as on a satellite.

1) Array on Satellites: It is possible to use an array on board a satellite and provide communication in a number of ways. For example, different frequencies may be allocated to beams covering different areas such that each area acts as a cell. This allows frequency reuse similar to the base-mobile system discussed previously. The major difference between this and the land-mobile system is the generation of beams covering different cells rather than having different base stations for different cells. An array mounted onboard the satellite provides the beam-generation facility. Depending upon the type of array system utilized, many scenarios are possible.

a) Fixed-shape beams: In a simplistic situation, beams of fixed shape and size may be generated to cover the area of interest, allowing normal handoff as the mobiles roam from one cell to another or as the beams move in a low orbit multiple satellite system covering different areas. Proposed fixed-beam antennas using multi-beam phased arrays for mobile communications and featuring frequency-reuse facility and flexibility of variable beam power allocation have been discussed in later chapters. Traditionally, phased arrays have been used in a feed network to control the beam coverage area. A typical system may consist of a high-gain large reflector antenna along with an array of feed elements placed in the focal plane of the reflector in a particular geometry able to generate a limited number of fixed-shape spot beams. A particular beam is selected by choosing a combination of feed elements. The steering of beams is achieved by controlling the phases of signals prior to the feed elements. Dual circuitry is employed for transmit and receive modes for the signal to flow in both directions. The capability of the system to generate multiple spot beams with independent power control and frequency use makes it attractive for mobile communications. The frequency scanning system considered uses an array of active antennas to provide a high gain beam with the capability to steer it at any user location using frequency-dependent inter-element phase shift. The same array is used for transmit as well as for the receive mode. In the transmit mode, the signal is supplied to antennas by an array of power amplifiers, whereas in the receive mode, the antennas feed the low noise amplifiers of the receiver. A comparison of the two systems indicates that the frequency scanning system outperforms the conventional multi-spot system in terms of capacity, required power per channel and payload mass per channel but requires a complex beam-forming network and mobile terminal location procedures.

b) Dynamic beams: A system using fixed-shape beams does not require knowledge of the traffic conditions, as does a system generating spot beams of varying shapes and sizes dictated by the positions of the mobiles (Fig. 11). For downlink transmission, this helps in transmitting the energy in the direction of mobiles by controlling the shape and coverage of the beams, which in turn reduces co channel interferences at the mobiles. This also helps reduce the transmitted power due to its directed nature of transmission, relaxing the need to generate large amounts of power onboard a satellite. To generate an arbitrarily shaped beam to be pointed at a desired location, the system architecture requires an advanced beam-forming network with independent

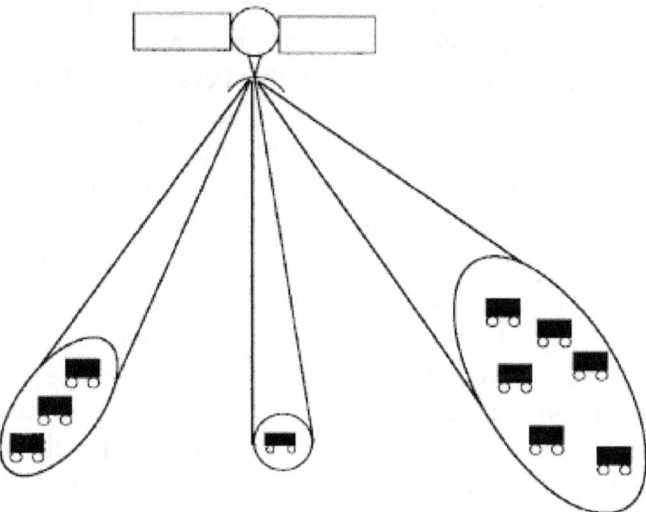

Fig 11:- Satellite systems generating spot beams of various shapes to cover cluster of mobiles.

beam-steering capability. Recent developments in monolithic MMIC technology, which allows fabrications of power amplifiers, phase-shifters, and low noise amplifiers directly coupled to radiating elements, provides multi-beam functionality using a very small space onboard the satellite. The technology enables one to fabricate a large number of active elements that may be independently controlled by digital hardware. This, along with digital signal processing, gives one the capability to generate a large number of independent beams of arbitrary shapes, which may be steered at any desired location. Such active elements with dedicated high-power and low-noise amplifiers have been used in recent system developments, including the GLOBALSTAR and IRIDIUM systems. Panels containing a large number of elements are mounted onboard the satellite, and these elements are used to generate many fixed-shape beams. These beams form co channel cells, with each cell supporting a number of communications channels. The system allows normal handoff when mobiles move out of a cell and does not use the knowledge of the positions of mobiles to adapt the beams. Each beam supports a group of mobiles.

c) Separate beam for each mobile: A satellite acts as a relay station between mobiles and the base station, with communication between the base station and the satellite being at a different frequency than that between mobiles and the satellite. It is envisaged that each mobile is tracked and the beam is pointed toward the desired mobile, with nulls in the directions of other mobiles operating at the same frequency. Different frequencies are used to communicate with the mobiles in close vicinity. Though the system as described uses different frequencies, the principle of operation remains the same for CDMA and TDMA systems. It should be noted that the direction-finding and beam forming algorithms in this case operate in a different environment than those operating for the base-mobile communications systems. Due to the distance involved between the satellite and the mobiles, the signals arriving from the mobiles appear more like point sources, which is not the case for the signal arriving at a base station due to spreading of signals caused by reflections in the vicinity of mobiles. The problem of multipath fading and delayed arrival is also not serious in this case.

2) Array on Mobiles: An array of antennas may be mounted on a mobile (vehicle) to communicate directly with a satellite, along with its control circuitry, to steer a beam toward the satellite. As the direction of the satellite with respect to the mobile is changing constantly due to vehicle movement, it requires constant tracking of the satellite and adjusting of the direction of the beam such that it points toward the satellite. Apart from this, the structure of the land-mobile antenna also needs to take into account the aesthetic aspects, which is not the case for base-station antennas. A number of studies have been reported in the literature covering a satellite communications system utilizing multiple antennas for land mobiles, which highlights these and other issues requiring consideration. Some are briefly mentioned here. The mobile satellite experiment studied uses a number of antennas mounted on a vehicle. A beam is electronically steered in the direction of a satellite using phase shifters. The system consists of many satellites and a vehicle communicates with the one that is in view at the time, requiring tracking of all the satellites and switching between the satellites as required. Electronic steering is not utilized for the German TV-SAT 2 system using multiple antennas. It employs four sets of pre-oriented antenna elements mounted on four faces of truncated pyramids. Fixed beams from each set are formed separately and are switched depending on the orientation of the mobile relative to the geostationary satellite used. The system is useful for larger vehicles, such as buses and trains. Separate elements are suggested that are alternately placed side by side in a planar configuration for transmit and receive mode using digital beam-forming techniques for tracking two satellites in multi-satellite systems. The system uses low-orbit satellites. A spherical array mount, useful for ships and aircraft employing digital beam forming to control beam direction, Test results on the characteristics and the suitability of an antenna array mounted on an aircraft to communicate directly with a satellite. An experimental development of a four-element antenna array to receive a 1.537-GHz pilot tone from the INMARSAT II F-4 geostationary satellite with a fixed position receiver shows that a significant improvement in link reliability can be produced using an adaptive array at a hand-held mobile compared to an Omni directional antenna. It indicates the possibility of using an array at hand-held mobiles communicating directly with a satellite, provided that the required signal processing could be incorporated somewhere.

Satellite-to-Satellite Communications

Inter-satellite communications will play an important role in mobile communications, particularly with the use of low-orbit satellites. The position of these satellites changes quickly, which causes a change in their relative directions. Arrays of antennas can play a major role in this situation by forming beams that always point in the direction of the desired satellite and have nulls in the direction of others operating in the same band. The array may be used for tracking the satellites as well as for canceling interferences arising from the transmission of other satellites. It may also provide protection against unfriendly jamming. An application of adaptive arrays to cancel interferences in a satellite-to- satellite communications system using direct-sequence spread spectrum. A constrained beam-forming algorithm, which uses the direction of the desired signal as well as an adaptive beam forming using the reference signal, is studied for this application. The study assumes that the direction of the desired satellite in this case may be obtained from the knowledge of the position of the satellite as it follows a well-defined trajectory.

Smart antenna system and its application in Low-earth-orbit satellite communication systems

Space-division-multiple-access is one of the solutions for increasing system capacity as well as improving system performance in wireless communications. Experimental studies have been reported, using a smart antenna system to exploit spatial diversity in low-earth-orbit (LEO) satellite personal communication services **(S-PCS).** The paper presents direction-of-arrival (DOA) estimation of satellite signals, variation of spatial signature and DOA with frequencies, elevation angle and motion, and diversity gain achieved by using an antenna array in line-of-sight (LOS) and shadowing scenarios. For typical environments, such as LOS and light or heavy shadowing cases, the experimental results reveal rich spatial diversity and high diversity gain when using a smart antenna system in LEO S-PCS. Low-earth-orbit (LEO) satellite communication systems at the L- and **S-** bands, such as Globalstar and Iridium, have been proposed to provide a global service. The satellite-earth communication channel depends on local features of the environment. Traditional low-gain and Omni-directional antennas employed in satellite communication systems are susceptible to multipath fading, which degrades reception when the signal path is blocked by buildings or shadowed by trees. By utilizing a phased array antenna, composed of multiple antenna elements, and applying advanced array signal-processing algorithms at the hand-held subscriber units or telephone booths of an LEO satellite system, we can combat multipath fading, reduce interference, increase power efficiency, obtain diversity gain, and improve system performance. A system that employs more than one radiating antenna element and utilizes active components in adapting the antenna's radiation patterns to improve some operating parameters of the system, such as CA, coverage and capacity, is called a smart antenna system. A mobile user with an M-element antenna array receives signals from a satellite at a certain elevation angle along the direct path, through the ground reflection path, and diffused multipath. Those signals come from different directions of arrival (DOA's) and can be estimated by direction- finding algorithms. A spatial signature **(SS)** is the response vector of an antenna array receiver to a typical satellite constellation and local environments. Studying the variation of **SS** and DOA due to different environments, carrier frequencies, and relative motion between mobile user and satellite, lead to an understanding of the performance of space division- multiple-access (SDMA) and beam forming schemes for different scenarios in LEO satellite communication systems. By exploiting spatial diversity among antenna elements, multipath fading problems can be mitigated or eliminated. In addition to the problem of multipath fading, another significant problem is the limited power capability of the mobile terminal. With the LEO satellite distance ranging from 500 to 2000km, the hand-set may not have sufficient power to directly up-link because of the large path loss. Furthermore, fast fading due to rapid environmental change may saturate the dynamic range of the receiver. An antenna array can obtain diversity gain and reduce the probability of a sudden power variation. The purpose of this study is to apply the smart antenna system and SDMA scheme to the land-mobile satellite propagation channel. A series of measurements have been conducted to demonstrate the rich spatial diversity and diversity gain by using an antenna array in mobile stations of a LEO satellite mobile communication system. The effects of environment types and satellite elevation angles have been studied. We present the concept of the smart antenna system, and a description of the experimental set-up and mobile user environments. The results from different scenarios are also presented. Demand for wireless communications has grown exponentially during the last few years. With such rapid growth, the most significant problem for wireless communications is how to increase channel capacity. In addition to the capacity problem, the dramatic increase in radio traffic aggravates existing difficulties, such as multipath fading, channel reuse among neighboring cells, near-far receiving problems, hand-off from one cell base station to another, limited battery life for pocket handsets etc. Different schemes, such as FDMA, TDMA and CDMA, have been proposed to increase the number of users in a fixed spectrum slot. Since different mobile users transmit from and receive at different spatial locations, in addition to frequency, time, and code diversities, there is a very rich spatial diversity that can be exploited to significantly increase the system capacity, as well as improve the system performance. However, this spatial diversity is not exhibited on a traditional single-antenna system, but rather requires the use of spatially separated multiple antennas or an antenna array. Therefore, any SDMA system must have an antenna array at a base station to exploit the spatial diversity among different users.

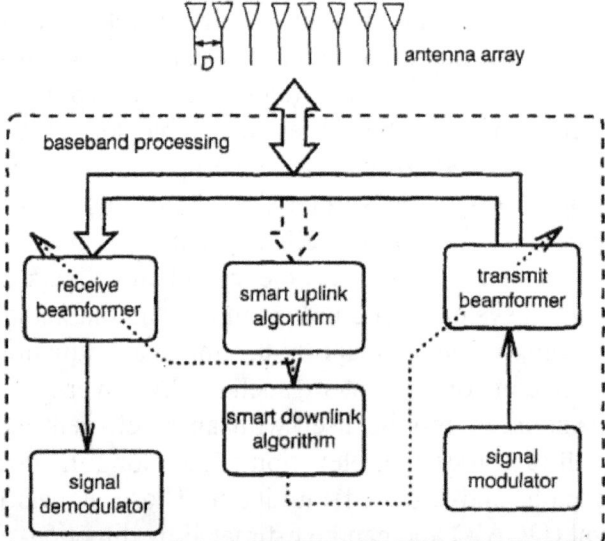

Fig: 12:- Smart antenna system base station

Fig.12 shows the basic implementation of a smart antenna system. The signal received by the antenna array is processed with advanced signal-processing algorithms, and the signal of interest is extracted from the received waves by exploiting the spatial diversity between the signal of interest and other signals, such as co-channel interference or background noise. Spatial diversity is actually the difference between the so-called spatial signatures of mobile terminals. Spatial diversity can also be exploited to generate transmitted beam forming patterns so that each signal is delivered to its desired location with minimum interference to other users.

Smart antennas are becoming one of the promising technologies to meet the rapidly increasing demands for more capacity of satellite communication systems. A main component in a smart antenna system is beam forming. Because of the limitations of analog beam forming, digital beam forming will be employed in future satellite communication systems. We evaluate the performance of various digital beam forming strategies proposed in the literature for satellite communications: 1) single fixed beam/single user 2) single fixed beam/multiple users 3) single adaptive beam/single user. Multiple criteria including coverage, system capacity, signal-to-interference-plus-noise ratio (SINR), and computation complexity are used to evaluate these satellite communication beam forming strategies. In particular, a Ka-band satellite communication system is used to address the various issues of these beam forming strategies. For the adaptive beam forming approach, sub array structure is used to obtain the weights of a large 2D antenna array, and a globally convergent recurrent neural network (RNN) is proposed to realize the adaptive beam forming algorithm in parallel. The new sub array-based neural beam forming algorithm can reduce the computation complexity greatly, and is more effective than the conventional least mean square (LMS) beam forming approach. It is shown that the single adaptive beam/single user approach has the highest system capacity. The demands for broadband services by satellite communications are growing rapidly. Future satellite systems will provide many new services such as high-speed internet access and broadband multimedia services. Because of the limited frequency spectrum for satellite communications, it is very important to improve the spectrum efficiency to enhance the system capacity. The smart antenna technology becomes one of the most promising approaches to improve the capacity of wireless communication systems, since a smart antenna can formulate multiple beams to realize the space division multiple accesses (SDMA) by dividing the service area into a large number of cells, and each cell is served by one beam. Currently many satellite communication systems use constellations to provide worldwide continuous services. However, the service areas are usually quite large to be covered by a single beam. In addition, the recent tremendous growth in communication industry has resulted in the crowding of the spectrum, and users in a service area may not be able to be served by a single beam. To accommodate more users, multiple beams can be used to increase the system capacity. A geo-stationary satellite may provide a number of spot beams at the same frequency, resulting in the frequency reuse by different beams and the increase of the system capacity. The basic idea of this SDMA approach is the generation of a large number of independently steered high-gain beams. Various satellite systems have incorporated this multiple beams capability into their systems. However, most of these satellite systems are based on the analog beam forming techniques. Analog beam forming has a serious restriction on the number of beams it can generate. More precisely, these analog beam forming techniques such as the Butler matrix must have orthogonal output beams and hence cannot fully exploit the advantage of SDMA. In contrast to analog beam forming, the digital beam forming technique can generate a large number of high-gain beams without degradation in signal-to-interference-plus-noise ratio (SINR). A digital beam forming technique can replace the complex hardware functions of a beam forming network by using software algorithms implemental on digital signal processors. This is particularly attractive to satellite communications since only a new suite of software has to be uploaded to the operating satellite if the beams require some modifications. Different digital beam forming strategies have been proposed for wireless communications. They can be categorized into two major classes: fixed digital beam forming and adaptive digital beam forming. For the fixed digital beam forming strategies the weights of the antenna array are designed in advance and kept unchanged during operations. For the adaptive digital beam forming strategies the weights are adjusted adaptively with the operating conditions. It is expected that adaptive beam forming algorithms can provide a further enhancement of system capacity because of their adaptive interference nulling capability. In addition, these two major classes can be further divided into four digital beam forming schemes, namely, 1) single fixed beam/single user, 2) single fixed beam/multiple users, 3) single adaptive beam/single user. While these four schemes are based on either fixed or adaptive beams, their difference lies in the use of a single beam for multiple users or a single user. Although these digital beam forming schemes have been proposed for SDMA in the literature, there has not been much work done in evaluating their effectiveness and practicalities for satellite communication applications. Many adaptive beam forming algorithms have been developed in the literature. Among them, minimizing the system output power subject to multiple linear constraints is a popular one, and a penalty function method can be used to overcome the weight jitter and high side lobe problems. But to be applicable for large arrays as in satellite communications, techniques such as the contiguous and overlapping sub arrays have to be considered. One new category of adaptive beam forming is neural beam forming where adaptive processes are carried out using neural networks (NN) for real-time parallel processing. Different NN models have been proposed in the literature. For example, the Hopfield NN (HNN) can realize the minimum mean-square error beam forming based on a neural optimization. Another approach uses a feed forward NN to approximate the input-output relationship for beam forming. Since the latter can be considered as nonlinear beam forming, we restrict our study to beam forming based on general beam forming techniques.

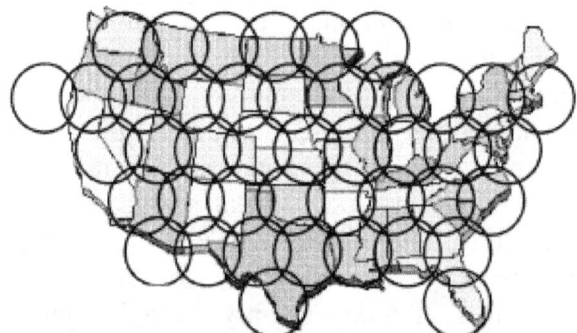

Fig.13:- Multi beam coverage of Ka band satellite communication

Some specifications of a Ka-band satellite communication system over the United States are given below. The Ka-band satellite system is used to investigate the effectiveness of these digital beam forming strategies. The coverage area is shown in Fig.13. There are 43 beams altogether. All beams have a coverage area of the same size. The specific parameters used in this study are given as follows:

- Half-power beam width (HPBW) of each beam: 1:0±.
- Angular separation of two adjacent beam centers: 0:9±.

- SINR: > 10 dB.
- Uplink frequency: 29.75 to 30 GHz, 250 MHz bandwidth.
- Downlink frequency: 19.95 to 20.2 GHz, 250 MHz bandwidth.
- Single uplink channel bandwidth: 2.7, 5.4, 13.5 MHz.
- Single downlink channel bandwidth: 60 MHz.
- Modulation scheme: BPSK (binary phase-shift keyed).

Digital Beam forming strategies for satellite communication systems

Digital beam forming can be carried out by using fixed digital beams or adaptive digital beams. For the fixed digital beam strategies, beams of fixed shapes and sizes are generated to cover the area of interest. A fixed beam is generated by a set of fixed weights. The weights are designed in advance and kept unchanged during operations. For the adaptive digital beam strategies, the beams vary dynamically according to the operating conditions. The weights are estimated by some adaptive beam forming algorithms. In this section, we describe the four possible digital beam forming strategies for satellite communications.

Single Fixed Beam/Single User Strategy
For the single fixed beam/single user strategy, each beam supports only one user, and each user has a different weight. This strategy may be considered as a non-null beam forming approach. While the weight of the beam is fixed, the beam is not reshaped to place nulls in the patterns of other co channel beams. Low side lobes are used to reduce the co channel interferences. In order to meet the system SIR/SINR requirements, the weights must be designed to produce very low side lobes. In the above Ka-band satellite communication system, the HPBW beams are circles. So we choose a rectangular array antenna to produce approximately circular fixed beams.

Single Fixed Beam/Multiple Users Strategy
The single fixed beam/multiple user's strategy is quite similar to the single fixed beam/single user approach. The weights are designed in advance and kept unchanged during operations. The main difference is the number of the users supported by each beam. There are multiple users in each beam for the single fixed beam/multiple user's scenario. Users in the same beam utilize the same weight while users in different beams use different weights.

Single Adaptive Beam/Single User Strategy
For the single adaptive beam/single user strategy, each beam is assigned to one user while beam forming is carried out using an adaptive algorithm. Each user is tracked, and deep nulls are formed along the directions of other users operating at the same frequency. Users in different beams use different sets of adaptive weights. To avoid grating lobes, the inter-element distance of the 2D array along x axis and y axis must be less than equal to 2. Conventional adaptive beam forming algorithms become impractical for such a large array. To reduce the computational complexity, a sub array approach is adopted. In addition, a globally convergent RNN is proposed here so that the multiple linearly constrained beam forming algorithm can be realized using parallel implementation.
Unlike the contiguous and overlapped sub array structures, we form sub arrays by uniformly decimating the elements of the phased array.

Use of Arrays in Transmit Mode

Antenna arrays have an equal role play in both receive and transmit modes. Their use in various configurations, as discussed previously, is applicable for both cases. Conceptually, the ideas of forming multiple beams to cover the cell site as well as pointing independent beams toward a cluster of mobiles, steering nulls toward co channel mobiles, and forming cells dynamically are equally useful for both the cases. However, the required signal-processing and hardware techniques to implement the array system for the two cases differ significantly. Conventionally, most of the investigations involving adaptive arrays have been for receiving antennas. The problem of transmitting co channel signals using an antenna array to several mobiles such that each mobile receives its desired signal with minimum cross talk with the other signals. It also proposes the formation of a desired transmitting beam pattern using the knowledge of propagation conditions acquired from the signals transmitted by each mobile for this purpose. It is suggested that the complex conjugate of the array weights, estimated during the reception by optimal combining, be used in transmit mode using different time slots for the two modes. It should be noted, however, that the optimal weights may change substantially, within a dwell time of 5 ms between the uplink and the downlink, causing a severe degradation in the performance of the system.

Performance Improvement Using an Array

An antenna array is able to improve the performance of a mobile communications system in a number of ways. It provides the capability to reduce co channel interferences and multipath fading, resulting in better quality of services, such as reduced BER and outage probability. Its capability to form multiple beams could be exploited to serve many users in parallel, resulting in an increased spectral efficiency. Its ability to adapt beam shapes to suit traffic conditions is useful in reducing the handoff rate, which may result in increased trunking efficiency. This section discusses the advantages of an array of antennas in a mobile communications system and improvements that are possible by using multiple antennas in a system rather than a single one. It provides references to experiments and studies where such improvements have been realized and highlights complexities associated with the implementation of such systems.

Reduction in Delay Spread and Multipath Fading

Delay spread is caused by multipath propagation where a desired signal arriving from different directions gets delayed due to the different travel distances involved. An array with the capability to form beams in certain directions and nulls in others is able to cancel some of these delayed arrivals in two ways. First, in the transmit mode, it focuses energy in the required direction, which helps to reduce multipath reflections causing a reduction in the delay spread. Second, in the receive mode, an antenna array provides compensation in multipath fading by diversity combining, by adding the signals belonging to different clusters after compensating for delays, and by canceling delayed signals arriving from directions other than that of the main signal.

1) Use of Diversity Combining

Diversity combining achieves a reduction in fading by increasing the signal level based upon the level of signal strength at different antennas whereas in multipath cancellation methods, it is achieved by adjusting the beam pattern to accommodate nulls in the direction of late arrivals, assuming them to be as interferences. For the latter case, a beam is pointed in the direction of the direct path or a path along which a major component of the signal arrives, causing a reduction in the energy received from other directions and thus reducing the components of multipath signal contributing to the receiver.

2) Combining Delayed Arrivals

A radio-wave originating from a source arrives at a distant point in clusters after getting scattered and reflected from objects along the way. This is particularly true in scenarios with large buildings and hills where delayed arrivals are well separated. One could use these clustered signals constructively by grouping them as per their delays compared to a signal available from the shortest path. Individual paths of these delayed signals may be resolved by exploiting their temporal or spatial structure.

The resolution of paths using temporal structures depends upon the bandwidth of the signal compared to the coherence bandwidth of the channel and increases as this bandwidth increases. In a CDMA system, the paths may be resolved provided their relative delays are more than the chip period. When these paths are well separated spatially, an antenna array may be used. This could be done, for example, by determining their directions. Spatial diversity combining similar to that used in the RAKE receiver may also be employed to combine signals arriving in multi-paths. The signals in each cluster may be separated by user specific information present in each signal, such as the frame identification number or the use of a known symbol in each frame. A simulation study reported uses the conjugate gradient method to adjust the array weights for compensating multipath fading in a land-mobile communication. The system uses TDMA, employs known symbols in each frame, and provides satisfactory performance using six elements in the presence of up to 30 scattered waves. In some situations, it might be possible to separate the signals in each cluster by forming multiple beams in the directions of each component of these clusters, with nulls pointing toward the other ones. Combining signals belonging to different users after compensating for delays not only leads to reduction in delay spread but also reduces co channel interference.

3) Nulling Delayed Arrivals

A reduction in delay spread using an array by nulling the delayed signals arriving from different directions has been studied in various literatures. The simulation study considers indoor radio channels, uses a PN sequence as a reference signal and the SMI algorithm to estimate the array weights, and concludes that using an adaptive array, a substantial reduction in delay spread is possible. Similar conclusions are reached using an experimental array of four elements mounted on a vehicle. It shows that using a CMA, the array is able to null the delayed arrival in a time-division-multiplexed channel. A frequency-hopping system may also be used for correcting degradation due to fading. The system is useful in frequency-selective fading, where the different frequencies fade differently. The frequency-hopping system is a spread spectrum system where the carrier frequency of transmission is changed in a predetermined manner, in contrast to a direct-sequence spread-spectrum system, where a pseudorandom sequence is used to spread the spectrum. An introduction of enough redundancy in such a system by coding the information before transmission may help the system to correct the loss of reception caused by fading at certain frequencies. A comparison of such systems with a system using antenna diversity of two antennas shows that the performance of the two systems is almost identical.

However, there is a cost associated with the coding method. It increases the rate of information due to added redundancy, requiring higher bandwidth. This in turn reduces the capacity of the system. Application of adaptive arrays for frequency-hopping communications systems is described in some papers whereas the use of adaptive arrays for direct-sequence spread-spectrum communications systems may be found.

Reduction in Co channel Interference

An antenna array has the property of spatial filtering, which may be exploited in transmitting as well as in receiving modes to reduce co channel interferences. In the transmitting mode, it can be used to focus radiated energy by forming a directive beam in a small area where a receiver is likely to be. This in turn means that there is less interference in other directions where the beam is not pointing. An analysis of a base station using multiple beams covering various mobiles indicates that co channel interference decreases as the number of beams increases. Co channel interference in transmit mode could be further reduced by forming specialized beams with nulls in the directions of other receivers. This scheme deliberately reduces transmitted energy in the direction of co channel receivers and requires knowledge of their positions. The reduction of co channel interference in the receive mode is a major strength of antenna arrays and has been reported widely. It does not require knowledge of the co channel interferences. If these were available, however, an array pattern might be synthesized with nulls in these directions. In general, an adaptive array requires some information about the desired signal, such as the direction of its source, a reference signal, or a signal that is correlated with the desired signal. In situations where the precise direction of the signal is known, interference cancellation may be achieved by solving a constrained beam-forming problem, whereas when this is not the case, it may be achieved by using a reference signal where the reference signal for a CDMA system is generated by code synchronization. A comparison of the two schemes for a satellite-to-satellite communications scenario when an approximate direction of the signal is known indicates that the constrained scheme cancels the interference faster than the reference generation scheme.

Arrays in Combination with Other Techniques

An antenna array may be used together with other methods to enhance the performance of a system by canceling the interference that might be present in directions other than that of the desired signal.

1) Spatial Diversity and Channel Coding: A combination of channel coding and spatial diversity (two antennas) for a TDMA indoor wireless system shows that the combined scheme provides a better BER than that without coding schemes. This reduces the effective data rate, however, and requires larger bandwidth for a given data rate to be transmitted. This in turn reduces the number of channels available. The problem may be alleviated by using extra antennas to increase space diversity.

2) Diversity combining and Interference Canceling:

An antenna array operating in a combined mode of diversity combining and interference canceling may be able to cancel directional interferences using some degree of freedom and achieve reduction in fading on the order of the remaining degree of freedom, thus improving the performance of the system as well as increasing its capacity. This increase depends upon the correlation of fading signals at different antennas. The degradation in the performance, however, is small for correlation up to 0.5.

3) Diversity combining and Adaptive Equalization:
The potential benefits of a system are consisting of diversity combining and adaptive equalization using a tapped delay line filter behind each antenna element. Diversity combining is useful for overcoming the flat fading in mobile communications, whereas the equalizers are normally used to reduce inter-symbol interferences in digital data transmission caused by delay spread. Thus, the combined structure offers an effective means to deal with the adverse effects of dispersive as well as flat fading. Many examples are provided to demonstrate the reduction in the average probability of the received bits for a QPSK modulated system using dual diversity and different types of equalizers under various environments. The importance of an antenna array using the tapped delay line filter to combat fading and to reduce co channel interference in wideband TDMA mobile communications channels is further emphasized to increase the data rate beyond about 2 Mb/s. The performance improvement of a QPSK modulated system employing an antenna array and a tapped delay line filter to combat multipath fading is a function of the number of taps.

Spectrum Efficiency and Capacity Improvement

Spectrum efficiency refers to the amount of traffic a given system with certain spectrum allocation could handle. An increase in the number of users of the mobile communications system without a loss of performance causes the spectrum efficiency to increase. Channel capacity refers to the maximum data rate a channel of given bandwidth could sustain. An improved channel capacity leads to more users of a specified data rate, implying better spectrum efficiency. In that sense, the two terms are used interchangeably in this paper following the convention used in most of the mobile communications literature. TDMA and CDMA result in an increase in channel capacity over the standard FDMA, allowing different time slots and different codes to be assigned to different users.

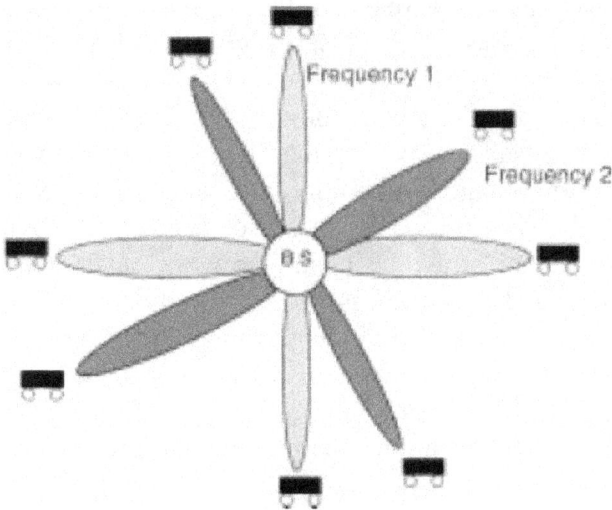

Fig 14:- Base station employing SDMA

Many studies have shown that this may be further improved by using multiple antennas and combining the signals received from them. First, the increased quality of service resulting from the reduced co channel interferences and reduced multipath fading, as discussed, may be traded to increase the number of users. Thus, the use of

an array results in an increase in channel capacity while the quality of the service provided by the system remains the same, that is, it is as good as that provided by a system using a single antenna. Second, an array may be used to create additional channels by forming multiple beams without any extra spectrum allocation, which results in potentially extra users and thus increases the spectrum efficiency. Fig. 8 shows a typical scenario of a base-station system employing an SDMA technique to serve many mobiles using only two frequencies. Fundamental limits on the maximum data rate and the capacity of a multiple antenna system in a Rayleigh fading environment concludes that two users with antennas each, using optimum transmitter/receiver pairs, may be able to establish up to channels, with each channel having about the same maximum data rate as that of a single channel. This and other studies on the impact of antenna diversity on the capacity of
a wireless communications system show the potential for a large capacity improvement in a communications system with the use of an antenna array. Salient features of some of these studies are now described to show the possible improvements in various configurations.

1) Multiple-Beam Antenna Array at the Base Station:

A base-station array in receive mode to locate the positions of mobiles in a cell and then transmitting in a multiplexed manner toward different clusters of mobiles one at a time using the same channel, the spectrum efficiency increases many times over and depends upon the number of elements in the array and the amount of scattering in the vicinity of the mobile. The study concludes that pointing beams with directed nulls toward other mobiles in a cell is a more efficient use of an array than reducing the cell size and the reuse distance. The use of a multiple-beam antenna array at the base station to improve spectral efficiency by resolving the angular distribution of mobiles shows that spectral efficiency increases as the number of beams increases. Formulas, which are helpful in predicting interference reduction and capacity increase provided by a switched-beam antenna system employed by a base station, also predict that the number of subscribers in a cell increases as the number of beams increases.

2) Antenna Array at the Base Station Employing CDMA

System: By using an array of antennas for a CDMA system, the number of mobiles a cell may be able to sustain increases manifold (on the order of the number of elements) for a given outage probability and BER. For example, at an outage probability of 0.01, the system capacity increases from 31 for a single antenna system to 115 for a five-element array. Using an array of seven elements increases the capacity to 155. The study reported is for both mobile-to-base and base-to-mobile links and derives expressions for the outage probability considering the effect of co channel interferences. The latter uses an RLS algorithm to estimate the weights of the array considered in simulation study and shows that capacity improvements on the order of 60% are possible using a four-element linear array.

3) Antenna Array at the Base Station Employing TDMA

System: Examples of a TDMA system employing antenna arrays reported uses the constant modulus adaptive algorithm for updating the weights of the array, whereas that another case study uses the SMI algorithm and considers the IS-54 system for its study. IS-54 is the standard for digital mobile radio in North America. It is for a

TDMA system where each user's slot contains a 28-b synchronization sequence along with 12-b for user identification and 260 data bits. The study uses the 28-b synchronization sequence as the reference signal and the user identification bits to make sure that the correct user's signal has been acquired. The study concludes that by using optimal combining at the base station with only two antennas, a frequency reuse factor of four is possible except in the worst cases at the boundaries of the cell. This could be further improved to a frequency reuse factor of three using dynamic channel assignment. Furthermore, it argues that frequency reuse in each cell is possible using four or five antennas. This is a substantial improvement compared to a frequency reuse factor of seven proposed for the current system.

4) Base-Station Array for Indoor-Mobile Communications:

Multipath fading is a severe problem for indoor-mobile communications. However, due to the slow speed of mobiles, the fading rate is much lower inside buildings than outside, typically on the order of a few hertz. This means that there is enough time to compute antenna weights for optimal combining using slow converging algorithms such as LMS, which offers the advantages of robustness and computational efficiency along with the possibility of its implementation using a single chip. Using theoretical analysis and computer simulations, it is argued that using optimal combining at a base station with antennas may lead to up to –fold capacity improvement in an indoor-mobile communications system. Furthermore, using a given number of antennas, a substantial capacity increase is also possible with a small increase in the SNR in a system. For example, a system using nine antennas may be able to achieve a six-fold capacity increase from a 10-dB increase in SNR per antenna. The replacement of the Omni directional base-station antenna with an adaptive array capable of forming multiple dynamically allocated beams steered toward the mobiles in a building complex reduces the delay spread substantially. This implies that by using an array, communications services may be extended to areas within buildings where delay spread is a serious problem, such as shopping centers, railway stations, and so on. The use of multiple antennas also increases the capacity of the network by allowing many users to operate at the same channels, which then may be separated by blind estimation methods. Depending upon the propagation conditions, a large improvement in capacity is possible.

BER Improvement

A consequence of a reduction in co channel interference and multipath fading by using an array in a mobile communications system to improve the communications quality is a reduction in BER and SER for a given SNR, or a reduction in required SNR for a given BER. A high-speed GMSK mobile communications system show that a four element adaptive array using the CMA is able to reduce the BER substantially compared to a single antenna system in a frequency-selective fading environment. The BER in a system normally reduces as the SNR is increased. These results, however, show that in an irreducible error rate environment, where no amount of SNR increase in a single antenna system would be able to reduce the BER, the system with an adaptive array is able to achieve large reductions. A simulation study employing a multi-beam system with co- channel interference capability indicates that for a QPSK modulated system, the SER decreases as the number of elements in the array increases, and in the presence of multipath arrivals, this decrease is better when there is a dominant path than when there are equal energy paths. A detailed analysis and computer study of the BER in a PSK demonstrates that using an array with optimal combining can achieve a large reduction in BER. Analysis and experimental results indicate that by optimally combining the signals received on multiple antennas at the base station, one is able to cancel the effect of other users in a flat Rayleigh fading environment to such an extent that the average probability of errors behaves as if the other users were not present. Computer simulations show that similar results hold for frequency-selective channels. An analysis of a PSK communications system shows that the use of an adaptive array reduces BER for such a system. An increase in the number of users in a cell normally causes an increase in the BER. The rate of this increase could be made lower by using an array compared to a single antenna system. The computer study uses the RLS algorithm to adjust the weights of an array and uses an expression for BER derived For an Omni directional antenna, it is given by:

$$P_e = Q * \sqrt{\frac{3G}{K(1+8\beta)-1}}$$

Whereas for multiple antennas it becomes:

$$P_e = Q * \sqrt{\frac{3GD}{K(1+8\beta)-1}}$$

Where is the standard -function showing the probability that for a zero mean, unit variance, Gaussian distributed random variable , is the processing gain of the CDMA system, is the number of uses in the cell, , and is the directivity of the beam of the multi antenna system. A comparison of the BER performance of a system using the conjugate gradient method and the RLS algorithm to adjust the weights of the array indicates the superiority of the former. The BER performance of the adaptive antenna system using the LMS algorithm is compared with the maximum entropy method and with a method based upon spatial discrete Fourier transform to show that both the latter methods provide better BER performance than the LMS algorithm.

Reduction in Outage Probability

Outage probability is the probability of a channel's being inoperative due to increased error rate in the received data. It may be caused by co channel interference in a mobile communications system. Using an array helps to reduce the outage probability by decreasing co channel interference. It decreases as the number of beams used by a base station for land mobiles increases in a multi-beam adaptive antenna system. The system analysis consists of calculating the outage probability considering one co channel cell as well as six co channel cells. The analysis shows that the reduction in outage probability is slightly less in the six-co channel-cell case compared to the one-co channel-cell case. Results on the outage probability of the system using an Omni directional base-station antenna analysis of the outage probability of a system with co channel interferences in a Nakagami-fading environment using diversity combining is done for various combining methods as a function of the number of elements in the array as well as the number of co channel interferences, and shows that the use of diversity combining reduces the outage probability. Results on the improvement in the outage probability using an array employing optimal combining concludes by theoretical analysis and computer simulation that the use of diversity combining results in a substantial reduction in the outage probability. A study of a CDMA system using a base-station antenna array derives equations for predicting the outage probability of the system for the uplink case, that is, by considering the signals arriving from mobiles. The study uses user-specific codes to determine the array response vector by correlating the array output with the code for the desired user. The estimated array response vector is then used for determining the array weight employing the optimal beam-forming technique. Numerical examples confirm that the outage probability increases by an increase in the number of users in a cell as well as by the presence of multipath. Furthermore, for a given outage probability, it shows that the use of an array allows an increase in the number of users in a cell, resulting in an increase in system capacity. The increase in capacity is on the order of the number of elements in an array.

Increase in Transmission Efficiency

Electronically steerable antennas are directive compared to fixed Omni directional antennas, that is, they have high gains in the direction where the beam is pointing. This fact may be useful in extending the range of a base station, resulting in a bigger cell size, or may be used to reduce the transmitted power of the mobiles. Experiment indicates that a reduction from 10 W to 250 mW is possible. This follows from the fact that by using a highly directive antenna, the base station may be able to pick a weaker signal within the cell than by using an Omni directional antenna. This in turn means that the mobile has to transmit less power and its battery will last longer, or it would be able to use a smaller battery, resulting in a smaller size and weight, which is important for hand-held mobiles. It is also advantageous to use an antenna array at the base station in transmit mode. In a single antenna system, all the power of the base station is transmitted by one antenna. However, when the base station uses an array of antennas and transmits the same amount of power as that of the single antenna system, the power transmitted by each antenna of the array is much lower compared to the case where the total power is transmitted by one antenna. Furthermore, for a given SNR at the mobile site, the base station using an array has to transmit less power compared to the single Omni directional antenna case due to the directive nature of the array. This further reduces the power transmitted by each antenna. These reductions in transmitted power level using an array allow the use of electronic components of lower power rating in the transmitting circuitry. This results in a lower system cost, leading to a more efficient transmission system.

Dynamic Channel Assignment

In mobile communications, channels are generally assigned in a fixed manner depending upon the position of a mobile and the available channels in the cell where the mobile is positioned. As a mobile crosses the cell boundary, a new channel is assigned. In this arrangement, the numbers of channels in a cell are normally fixed. The use of an array provides an opportunity to change the cell boundary and thus to allocate the number of channels in each cell as the demand changes due to changed traffic situations. This provides the means whereby a mobile or group of mobiles may be tracked as it moves and the cell boundary may be adjusted to suit this group. Dynamic channel assignment is also possible in a fixed cell boundary system and may be able to reduce the frequency reuse factor up to a point where frequency reuse in each cell might be possible. There may be situations when it is not possible to reduce co channel interferences in certain channels, and the call may have to be dropped due to large BER caused by strong interferences. Such a situation may arise when a desired mobile is close to the cell boundary and the co channel mobiles are near the desired mobile's base station. This could be avoided by dynamic channel assignment, whereby the channel of a user is changed when the interference is above a certain level.

Reduction in Handoff Rate

When the number of mobiles in a cell exceeds its capacity, cell splitting is used to create new cells, each with its own base station and new frequency assignment. A consequence of this is an increased handoff due to reduced cell size. This may be reduced using an array of antennas. Instead of cell splitting, the capacity is increased by creating independent beams using more antennas. Each beam is adapted or adjusted as the mobile locations change. The beam follows a cluster of mobiles or a single mobile, as the case may be, and no handoff is necessary as long as the mobiles served by different beams using the same frequency do not cross each other. A comparison of the SDMA method with other schemes indicates that the former offers many practical advantages over the other methods of capacity increase.

Reduction in Cross Talks

Cross talks may be caused by unknown propagation conditions when an array is transmitting multiple co channel signals to several receivers. Adaptive transmitters based upon the feedback obtained from probing the mobiles could help eliminate this problem. The mechanism works by transmitting a probing signal periodically. The received feedback from mobiles is used to identify the propagation conditions, and this information is then incorporated into the beam-forming mechanism.

Cost-Effective Implementation

The use of multiple antennas on direct broadcast satellites leads to a system that is cost effective due to its light weight, reduced power requirements, flexibility and robustness of antenna design, ease of pattern control, and use of more solid-state devices. It also provides a larger aperture and mechanism for error compensation by signal processing.

Angular Spreading and Its Impact on Performance

Angular spreading refers to a situation where a transmitted signal gets scattered in the vicinity of the source and a signal arrives at a receiver within a range of angles. In a base-mobile communications system, where the base station antenna is normally high away from the ground and a mobile is close to the ground, the angular dispersion is more pronounced in the vicinity of the mobile and arrives at the base station with an angular distribution. The range of angles becomes smaller as the distance between a mobile and the base increases. Experimental results indicate that an angular spreading of about 3 results from a distance of 1 km. Various models of the scattering situation have been reported in the literature, assuming the multipath signals to be distributed uniformly within the spread angle as well as with the Gaussian distribution. Selection of the distribution function, however, does not appear to be critical as long as the spread is small around the nominal direction. A dispersion of the radio environment results in the distortion of the perceived antenna side-lobe levels at the base station. As well as an increase in correlation of fading at different elements of the array both of which affect the performance of the system. The problem of fading correlations is studied is shown by deriving the relationships between the angle of arrival, beam-width, and correlation of fading that a larger antenna spacing is required to reduce the correlation when the angle of arrival approaches parallel to the array, resulting in a reduction of the beam-width. An increased correlation of fading between various elements greater than 0.8 causes signals at all of the antennas to fade away simultaneously, rendering an array in the maximal ratio combining mode ineffective against fading correction. The array, however, is able to suppress interferences, as independent fading is not required for interference suppression. A detailed investigation of the effect of fading correlation on the performance of adaptive arrays to combat fading is carried out concludes that correlation of up to 0.5 causes little degradation, but a higher correlation decreases performance significantly. Transmission strategies in a multi-hop packet radio network normally are concerned with designing routing algorithms for sending packets from the source to the destination. A study presented in indicates that fading affects the performance of a multi-hop system using routing algorithms. This, however, does not seem to be the case for a single-hop system.

Cost, Complexity, and Network Implication

It follows from the discussion in previous sections that a system using an adaptive array to improve the performance of mobile communications requires estimation and optimization of numerous time-varying system parameters in a dynamically changing environment. Though the specific parameters would depend upon the type of system under consideration and the mode of array processing incorporated in the system, it is clear that implementation of such a system would require a complex network and system architecture. Implementation of schemes such as dynamic handoff, dynamic channel assignment, dynamic beam shaping to incorporate clusters of mobiles, dynamic nulling of co channel interferences, and procurement of the knowledge of the desired mobile—in the form of a reference signal, its direction, or the array response vector associated with the mobile—for the purpose of beam forming would require a complex control structure. Then there is a question of the time required to update the system parameters. The system requires quick updates on the positions of fast moving mobiles, whereas its response time is limited by the time required by DOA estimation and tracking schemes to update the positions of mobiles and by the beam-forming algorithms to converge to a satisfactory level. Though some of these algorithms may be implemented in parallel to increase the signal processing power, this adds to the cost of the system. The system cost includes not only the cost of building hardware to implement the control structure but the cost of building active antennas along with the required phase shifters, high-power amplifiers, and circuitry associated to drive these components. Most of these components are excessively expensive at present and are not required for a system without SDMA. This is currently a disadvantage of using an adaptive array in mobile communications. Apart from the cost of implementation, many reliability questions also need careful consideration. Some network architecture and traffic-implication issues of mobile communications in general are considered. However, there is no consideration of the complications arising from the use of an array in such systems.

Feasibility of Antenna Array Systems

Array processing involves manipulation of signals induced on various antenna elements. Its capabilities of steering nulls to reduce co channel interferences and pointing independent beams toward various mobiles, as well as its ability to provide estimates of directions of radiating sources, make it attractive to a mobile communications system designer. Array processing is expected to play an important role in fulfilling the increased demands of various mobile communications services. Part I of this paper showed how an array could be utilized in different configurations to improve the performance of mobile communications systems, with references to various studies where feasibility of an array system for mobile communications is considered. This paper provides a comprehensive and detailed treatment of different beam-forming schemes, adaptive algorithms to adjust the required weighting on antennas, direction-of-arrival estimation methods—including their performance comparison—and effects of errors on the performance of an array system, as well as schemes to alleviate them.

1.8 Conclusion

This Chapter provided an overview of the potential benefits and challenges of applying smart-antenna technology to mobile communication systems. The basic advantages and disadvantages of two different approaches to smart antennas *(a)* SDMA/SFIR; *(b)* switched beam/direction finding/optimum combining), as they are presented in open literature, were highlighted. Then the importance of specific characteristics of the radio channel under different operational environments (e.g. large-small cells, indoor-outdoor) and interference scenarios (CDMA-TDMA) were discussed. The realization of comparable up and down link gains along and the practical implementation of a fully integrated smart antenna system seem to be the two most challenging issues facing this technology at the moment. Although more efficient implementations will almost certainly be needed for future systems, partial implementations make the application of smart antennas to current generation systems possible. Furthermore, by the time that future generation systems are ready fully to support smart antennas, the development of RF and DSP technology will have reached such a level that complex implementations will be possible. Exploiting different characteristics of smart antennas can lead to several operational benefits for a communication system. These benefits have been discussed in the paper, and are summarized below:

1. coverage extension,
2. increased capacity
3. efficient power control/smart handover
4. support of value added services (better signal quality, Higher data rates, user location)
5. optimum/smart system planning
6. Reduced transmit power
7. Smart link budget balancing

Communication systems will exploit different advantages or mixtures of advantages offered by smart antennas depending on the maturity of the underlying system. Initially, for example, costs can be reduced by exploiting the range extension capabilities of smart antennas. Then, where there is a demand for increased capacity, costs can be further decreased by avoiding extensive use of small cells and instead exploiting the capability of smart antennas to increase capacity. Finally, more advanced systems (3rd generation) will be able to benefit from smart antenna systems, but it is almost certain that more sophisticated space/time filtering approaches will be necessary, especially as these systems become mature.

CHAPTER 2

Basics of an Adaptive Antenna System

The limiting factor on the capacity of a cellular mobile system is interference from co-channels mobiles in neighboring cells. Adaptive antenna technology can be used to overcome this interference by intelligent combination of the signals at multiple antenna elements.

2.1 Basic concept

In a phased array, a set of antenna elements are arranged in a space and the output of each element is multiplied by a complex weight and combined by a summing as shown in figure (2.1).

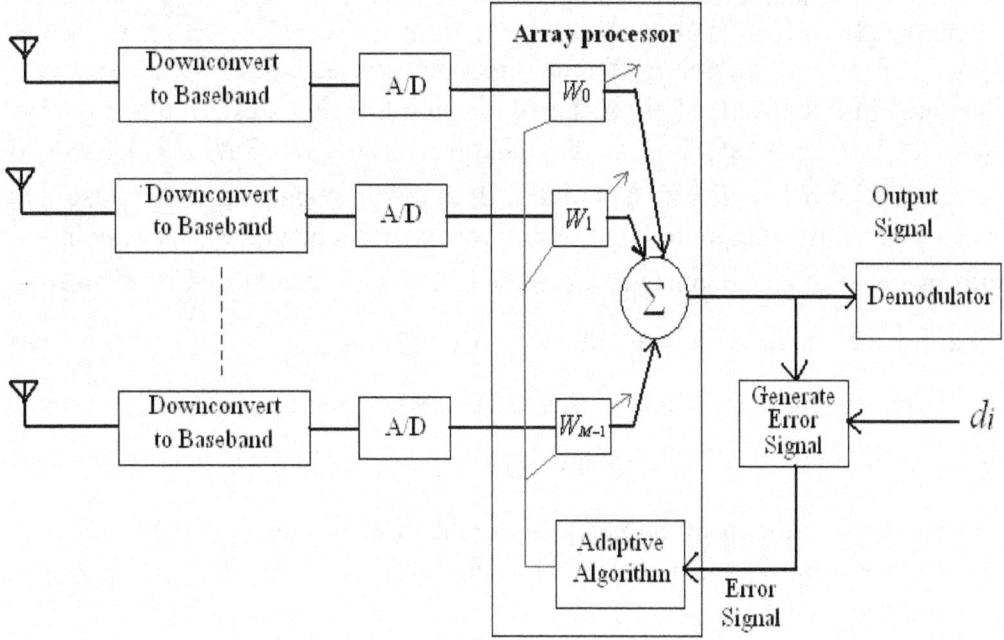

Figure 2.1:- An adaptive array structure. Note that d_i represents an estimate or replica of the desired signal for the ith user at the array output.

The complete array can be regarded as an antenna in its own right, with a new output y. The radiation patterns of the individual elements are summed with phases and amplitudes depending on the both the weights applied and their positions in space this yields a new combined pattern. If the weights are allowed vary in time, the array becomes an adaptive array, and it can be exploited to improve the performance of a mobile communication system by choosing the weights so as to optimize so measures of system performance. Typically this would be done by estimating the desired weights using a digital signal processor (DSP) and applying them in complex base band to sampled versions of the signals from each of element.

2.2 Degree of freedom

The gain and phase applied to signals derived from each element may be thought of as a single complex quantity, referred to as the weighting applied to the signals. If there is only one element, no amount of weighting can change the pattern of that antenna. However, with two elements, when changing the weighting of one element relative to the other, the pattern may be adjusted to the desired value at one place, that is, you can place one minima or one maxima anywhere in the pattern. Similarly with three elements, two positions may be specified, and so on. Thus, with an L element array you can specify L-1 positions. These may be one maxima in the direction of the desired signal and L-2 minima's (nulls) in the directions of unwanted interferences. This flexibility of an L element array to be able to fix the pattern at L-1 places is known as the degree of freedom of the array.

2.3 Example of Adaptive Antenna Processing

Let us now consider a simple example to illustrate the existence and calculation of the set of weights which will cause a signal from a desired direction to be accepted while"interference" from a different direction is rejected [8]. Such an example is illustrated in figure (2.2). Let the signal arriving from the desired direction $\theta = 0^0$ be called the "pilot" signal $p_k = P \sin k\omega_0$, and let the noise, $n_k = N \sin k\omega_0$ be incident to the receiving array at $\theta = \pi/6$ radians. Both the pilot signal and the noise signal are assumed for this example to be at exactly the same frequency ω_0. At a point in space midway between the antenna array elements, the signal and noise are assumed to be in phase. In the example shown , there are two identical omnis spaced $\dfrac{\lambda_0}{2}$ apart .The signals received at each element are fed to two variable weights one weight being preceded by a quarter –wave time delay of $\dfrac{\pi T}{2\omega_0}$ The four weighted signals are then summed to form the array output . The problem in figure (2.2) is to obtain a set of weights to accept p_k and reject n_k. Note that with any set of nonzero weights, the output is of the form $\qquad A \sin\left(k\omega_0 + \phi\right)$

And a number of solutions exist, which will make the output be p_k. However the output of the array must be independent of amplitude and phase of the noise if the array is to be regarded as rejecting the noise. Satisfaction of this constraint leads to a unique set of weights determined as follows. The array output due to the pilot signal is

$$P\left[\left(w_1 + w_3\right)\sin k\omega_0' + \left(w_2 + w_4\right)\sin\left(k\omega_0 - \frac{\pi}{2}\right)\right] \qquad (2.1)$$

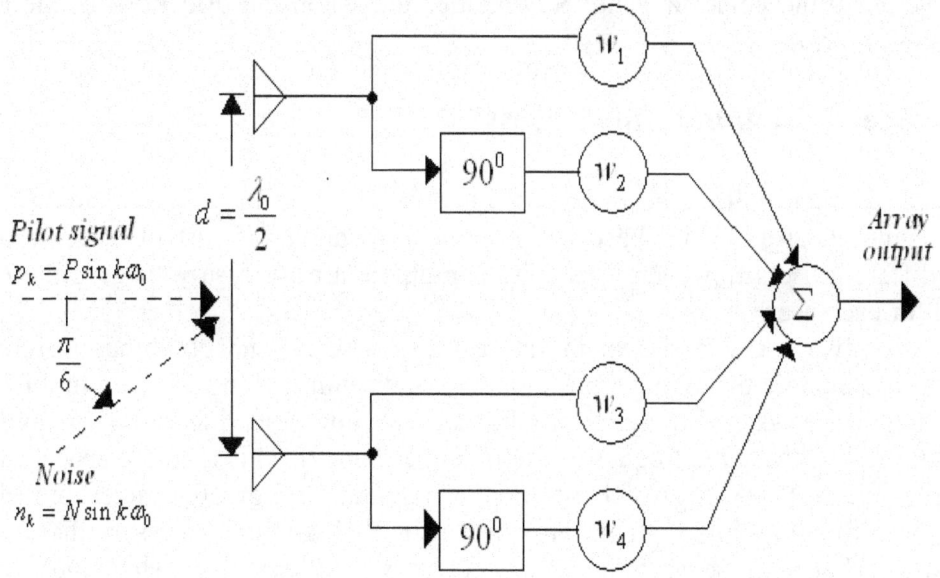

Figure 2.2:-Pilot signal arriving at $\theta = 0^0$ and noise arriving at $\theta = \pi/6$

For this output to equal the desired output $p_k = P \sin k\omega_0$ (which is the pilot signal itself) it is necessary that

$$w_1 + w_3 = 1 \tag{2.2}$$
$$w_2 + w_4 = 0$$

With respect to the midpoint between the antenna elements, the relative time delays of the noise at the two elements are $\pm(d/2c)\sin(\pi/6) = \pm d/4c = \pm \pi T/4\omega_0$ seconds, which corresponds to phase shifts of $\pm \pi/4$ at frequency ω_0. The array output due to the incident noise at $\theta = \pi/6$ is then

$$N\left[w_1 \sin\left(k\omega_0 - \frac{\pi}{4} \right) + w_2 \sin\left(k\omega_0 - \frac{3\pi}{4} \right) + w_3 \sin\left(k\omega_0 + \frac{\pi}{4} \right) + w_4 \sin\left(k\omega_0 - \frac{\pi}{4} \right) \right]$$

$$\tag{2.3}$$

For this response to equal to zero, it is necessary that

$$w_1 + w_4 = 0 \tag{2.4}$$
$$w_2 - w_3 = 0$$

Thus the set of weights that satisfies the signal and noises response requirements can be found by solving (2.2) and (2.4) simultaneously. The solution is

$$w_1 = \frac{1}{2}, w_2 = \frac{1}{2}, w_3 = \frac{1}{2}, w_4 = -\frac{1}{2} \tag{2.5}$$

With these weights, the array will have the desired properties in that it will accept signal from the desired direction, while rejecting noise which is at the same frequency ω_0 as the signal, but arriving from a different direction than that of the signal point.

The foregoing method of calculating the weights is more illustrative than practical. This method is usable when there is only a small number of directional noises sources, when the noises are monochromatic, and when the directions of the noises are known a priori.

2.4 Space division multiple Access

In space division multiple access technique a multiple beams are generated pointing towards mobile users. At the base station same array antenna is used to receive signals from different users however a separate combiners are necessary to generate output signal for each user.

A case for two users is shown in figure (2.3). The second combiner weights are chosen to eliminate the signal from mobile 1 and retain the signal from mobile 2. In this situation the system would be simultaneously communicating with two mobiles in the same cell on the same frequency / time /code channel. This is called space division multiple access (SDMA) which offers the potential for greatly increasing system capacity in future mobile systems. However using SDMA it is necessary that mobiles should have angular separation. Clearly there would be times when the multiple beams produced by base station overlap, making it impossible to separate the mobiles completely. Perhaps even more importantly, the scattering nature of the propagation channel will cause the signals received from the mobile to be broadened in arrival angle, making them overlap even if the mobiles have some angular separation. Thus the capacity of an SDMA system is limited by the capabilities of the adaptive array and by the characteristics of the channel. By using an array of antennas for a CDMA system, the number of mobiles a cell may be able to sustain increases manifold (on the order of number of elements) for a given outage probability and BER . For example, at an outage probability of 0.01, the system capacity increases from 31 for a signal antenna system to 115 for a five element array.

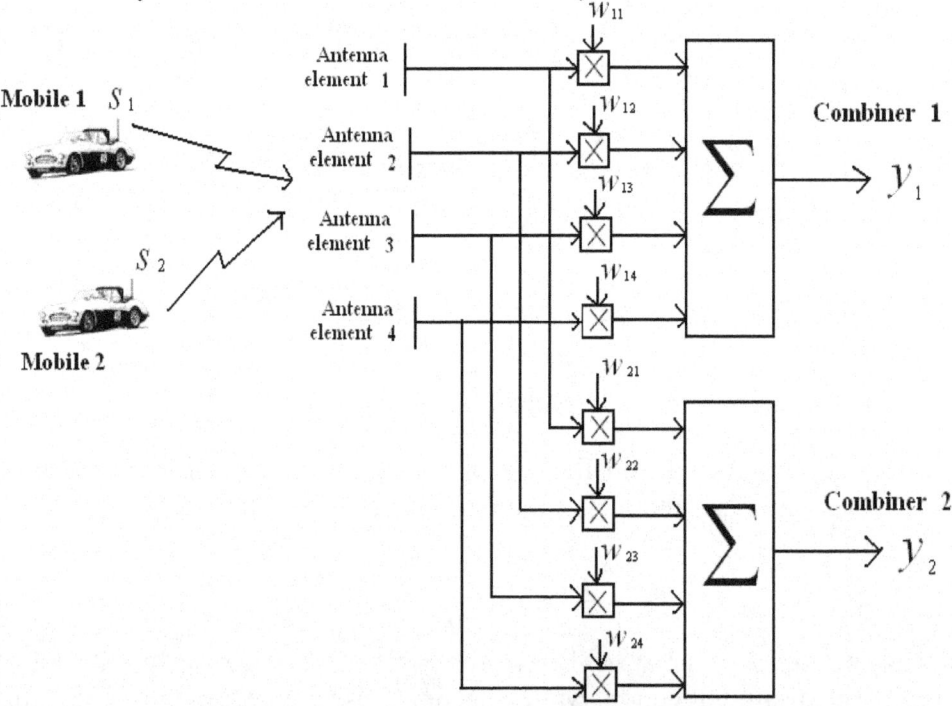

Figure 2.3:-Dual combining for two-channel SDMA

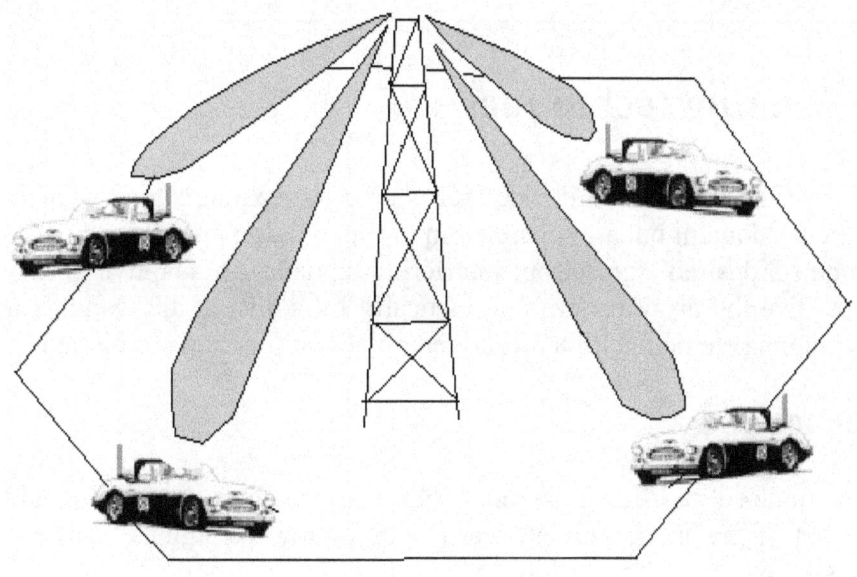

Figure 2.4:-Space division multiple access

Implementation of SDMA requires measure changes to the base station and is difficult to implement with systems for which SDMA was not originally foreseen.

2.5 Limitations of Smart Antennas

1) If an antenna consists of M elements in the array then, the hardware/ software requirements increases as M demodulators are required for each user.

2) The M receivers must be accurately synchronized in time to provide effective performance.

3) The computational complexity of array processing algorithms can be very large.

4) The array size will be constrained by the available space for a base station. Usually, the spacing of antenna elements varies from one half to tens of RF carrier wavelengths.

5) Practical antenna arrays may be adversely affected by channel modeling errors, calibration errors, phase drift and noise which are correlated between antennas.

CHAPTER 3

Beam forming techniques

An adaptive beam-former is a device that is able to separate signals which have the same frequency content but are separated in the spatial domain. This provides a means for separating a desired signal from interfering signals. An adaptive beam-former is able to optimize the array pattern automatically by adjusting the weights associated with each antenna element until a prescribed objective function is satisfied.

3.1 Beam forming

The signals induced in different elements of an antenna array are combined to form a single out put of the array. This process of combining the signals from the different elements is known as beam forming which describes the basic characteristics of an antenna.

To generate a steerable beam from a phased array it is necessary to accomplish three main functions:-

(a) Co-phase the signals arriving at the elements of the array from the desired direction.

(b) Apply amplitude weighting to control the spatial sidelobe structure.

(c) Sum the weighted co-phased signals to produce a wanted beam.

since a wave front arriving from a direction other than bore sight will reach different elements of the array at different times depending on their position in the array, the most logical method of co-phasing would be a to introduce variable time-delays into the elements outputs to compensate for these differential delays. Since the co-phased direction would then be independent of frequency, this method is preferred for arrays which must operate over instantaneous bandwidths, which are a significant fraction of the centre frequency. The range of time-delay T required will be a simple function of the largest array dimension, D, and the maximum desired scan angle A.

$$T = \frac{D}{c}\sin(A)$$

Where, c is the electromagnetic speed of propagation.

3.1.1 Analog Beam forming

An antenna array consisting of a number of antenna elements, the outputs of which are combined via amplitude and phase controlled network, in order to form a desired antenna beam. It is possible to perform analog beam forming at the RF stage, using phase shifters and amplifiers, however, the high specifications required of these devices renders them costly. An alternative solution is to down-convert the RF signal to an Intermediate Frequency (IF) and to perform the beam forming at the IF stage .The disadvantage of this technique is that each antenna must have its own RF-to-IF receiver. Multiple beam formers must be used to form multiple beams, resulting in the distribution of the signal energy across all the formed beams. The output SNR is thus reduced, when the lower signal energy of the beams is combined with the increased noise injected by the increased number of RF and IF stages. In an analog beam-former the magnitude of the weights can never be greater than unity, as that will imply the existence of an amplifier.

3.1.2 Digital Beam forming

The digital beam former requires the elements signals to be available in digital form, and since at any instant in time the signal vector carries both amplitude and phase information, each digital sample must encapsulate this information. The conventional way of transforming the signal to digital form is shown schematically in figure (3.1).

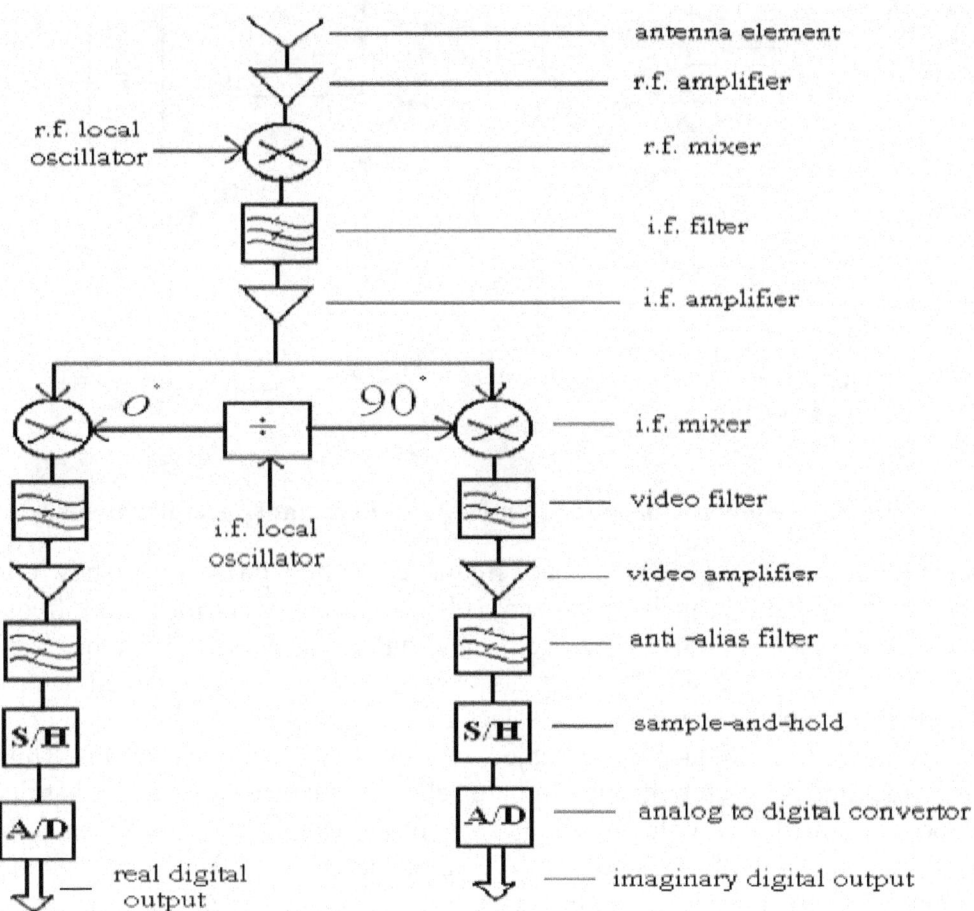

Figure 3.1:- Conventional receiving chain

In the conventional digital beam former, the signals are then multiplied by the appropriate set of beam forming coefficients and summed. The quadrature signal components can be treated as complex sample, and the coefficients complex coefficient, so the multiplication and summation are also complex.

In digital beam forming the weights are mere numbers and they can be greater than 1 and the beam pattern has no physical meaning, unlike its analog counter part. It is important to note that the sidelobe can exceed the mainlobe without interfering with the desired constraints.

3.1.3 Frequency-Domain Beam Forming

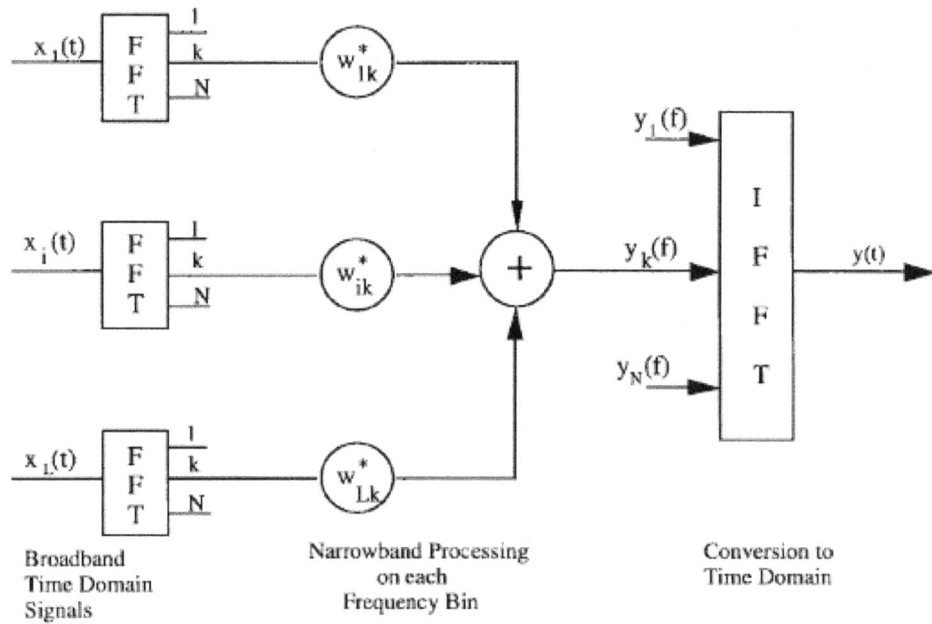

Fig 3.2:- element space frequency domain beam space processor

A general structure of the element-space frequency domain processor is shown in Fig. 3, where broad-band signals from each element are transformed into frequency domain using the FFT and each frequency bin is processed by a narrow-band processor structure. The weighted signals from all elements are summed to produce an output at each bin. The weights are selected by independently minimizing the mean output power at each frequency bin subject to steering-direction constraints. Thus, the weights required for each frequency bin are selected independently, and this selection may be performed in parallel, leading to a faster weight update. When adaptive algorithms such as the LMS algorithm (discussed later) is used for weight update, a different step size may be used for each bin, leading to faster convergence. The performance of the time- and frequency-domain processors is the same only when the signals in different frequency bins are independent. This independence assumption is mostly made in the study of frequency-domain beam forming. When this assumption does not hold, the frequency-domain beam former may be suboptimal. A study of the frequency-domain algorithm for coherent signals indicates that the frequency-domain method is insensitive to the sampling rate and may be able to reduce the effects of element malfunctioning on the beam pattern. Due to its modular parallel structure, beam forming in the frequency domain is well suited for VLSI implementation and is less sensitive to the coefficient quantization.

3.1.4 Broad-Band Beam Forming

As the signal bandwidth increases the performances of the narrow band beam-former deteriorates. For processing broad band signals a TDL structure shows in the figure.3.3 normally used. A lattice structure consists of J simple lattice filter sometimes is also used offering some processing advantages The steering delays in front of each element in Fig. 3.3 are pure time delays and are used to steer the array in a given look direction (ϕ_0, θ_0). If $\tau_L(\phi_0, \theta_0)$ denotes the time taken by the plane wave arriving from direction (ϕ_0, θ_0) and measured from the reference point to the lth element, then the steering delay $T_L(\phi_0, \theta_0)$ may be selected using $T_L(\phi_0, \theta_0) = T_0 + \tau_L(\phi_0, \theta_0)$ where T_0 is a bulk delay such that $T_L(\phi_0, \theta_0) > 0$ for all. If s(t)denote the signal induced, on an element present at the center of the coordinate system, due to a broad-band source of power density s(f)then the output of the lth sensor pre-steered in (ϕ_0, θ_0) is given by : $x_l(t) = s(t + \tau_L(\phi, \theta) - T_L(\phi_0, \theta_0))$

For a source in (ϕ_0, θ_0) it becomes $x_l(t) = s(t - T_0) \ldots\ldots s = 1, 2 \ldots, L$ yielding identical waveforms after pre-steering delays. The TDL structure shown in the figure following the steering delay on each channel is a FIR filter. The coefficients of these filters are constrained to specify the frequency response in the look direction. It should be noted that these coefficients are real compared to the complex weights of the narrow-band processor. Let w is defined by $w = [w_1, w_2, \ldots\ldots, w_J]^T$.

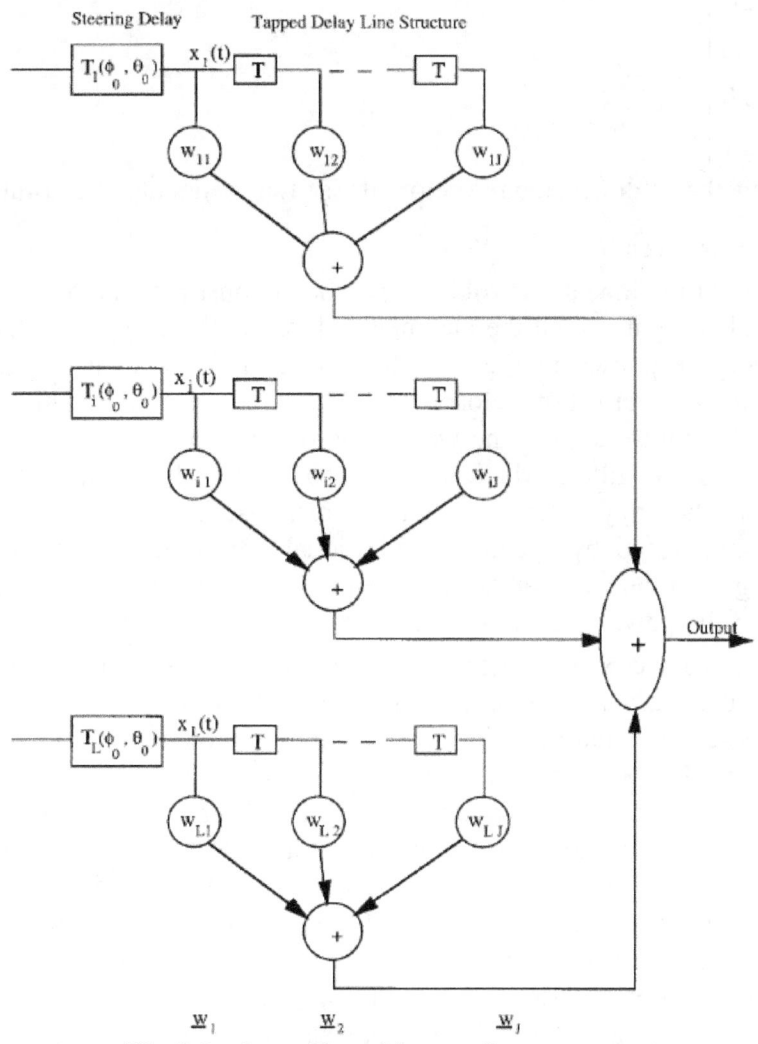

Fig 3.3:- broadband beam-former structure

Denote LJ coefficients of the filter structure with w_m denoting the L th coefficients after M-1 th tap The mean out power of the beam former for a given w is given by $P(w) = w^T R w$ where the $LJ \times LJ$ dimensional real matrix R denotes the array correlation matrix, with its elements representing the correlation between various tap outputs. The correlation between the outputs of the $l-1$th tap on the m th channel and the $k-1$th tap on the th channel is given by:

$(R_{m,n})_{l.k} = \rho[(m-n)T + T_l(\phi_0,\theta_0) - T_k(\phi_0,\theta_0) + \tau_k(\phi,\theta) - \tau_l(\phi,\theta)]$ with $\rho(t)$ denoting the correlation function. $\rho(t) = E[s(t)s(t+\tau)]$. It is related to the spectrum of the signal by the Fourier transform, that is $\rho(t) = \int_{-\infty}^{\infty} S(f)e^{j2\pi f\tau}df$

Thus, from the knowledge of the spectra of sources and their DOA's, the correlation matrix may be calculated. In practice, this can also be estimated by measuring signals at the output of various taps. In situations where one is interested in finding coefficients such that the beam former cancels the directional interferences and has the specified response in the look direction, the following beam-forming problem is normally considered: minimize w $w^T Rw$ subject to $C^T w = F$

Where F is a J dimensional vector that specifies the frequency response in the look direction and C is a LJ*J constraint matrix For a point constraint in the look direction

$$C = \begin{bmatrix} 1................0 \\ 0.1.............0 \\ 0.0.1..........0 \\ 0.0.0.0.0.0.1 \end{bmatrix}$$

With 1 denoting the L dimensional vector of 1s. Let w denote the solution of the above problem. It is given by $\hat{w} = R^{-1}C(C^T R^{-1}C)^{-1}F$

The point-constraint minimization problem specifies J constraints on the weights such that the sum of L weights on all the channels before the th delay is equal to F_j. For all pass frequency responses in the look direction, all but one F_j, j=1,2,.....J are selected to be equal to zero. For j close to J+1/2 , F_j is taken to be unity. Thus, the constraints specify that the sum of the weights across the array is zero except for one near the middle of the filter, which is equal to unity. The implication of these constraints is that the array pattern has a unity response in the look direction. This pattern can be broadened by specifying additional constraints, such as derivative constraints along with the constraints discussed above. The derivative constraints set the derivatives of the power pattern with respect to θ and ϕ equal to zero in the look direction. The higher the order of derivatives, that is, the first order, second orders, etc., the broader the beam in the look direction normally becomes. A broader beam is useful when the actual signal direction and the known direction of the signal is not precisely the same. In such situations the processor with the point constraint in the known direction of the signal would cancel the desired signal as if it were interference. The other directional constraints to improve the performance of the beam former in the presence of the look-directional constraints include multiple linear constraints and inequality constraints. A set of non-directional constraints to improve the performance of the beam former under look-direction errors is discussed are referred to as correlation constraints, which use the known characteristics of the desired signal to estimate an LJ dimensional correlation vector r_d between the desired signal and the array signal vector. The beam-forming problem using these constraints becomes minimize $w^T Rw$ subject to $r_d^T w = \rho$ where ρ_0 is a scalar constant that specifies the correlation between the desired signal and the array output. Application of broad-band beam-forming structures using TDL filters to mobile communications has been considered to overcome multipath fading and large delay spread in a TDMA as well as a CDMA system.

3.2 Pattern Adaptivity in Arrays

There are two main architectures for implementing adaptive digital beam forming.
1) Element space beam forming
2) Beam space beam forming

3.2.1 Element space beam forming

In this method, as shown in figure (3.4) signals arriving from each of the N elements are weighted and summed to produce the desired output. This allows generation of multiple simultaneous beams making full use of the degrees of the freedom for adaptive interference rejection. The beam controller accepts buffered samples from all the elements, together with information concerning the channel alignment coefficient, the required pointing directions of the beams, and any other beam constraints. The adaptive algorithm then computes the weights which are used by the beam former.

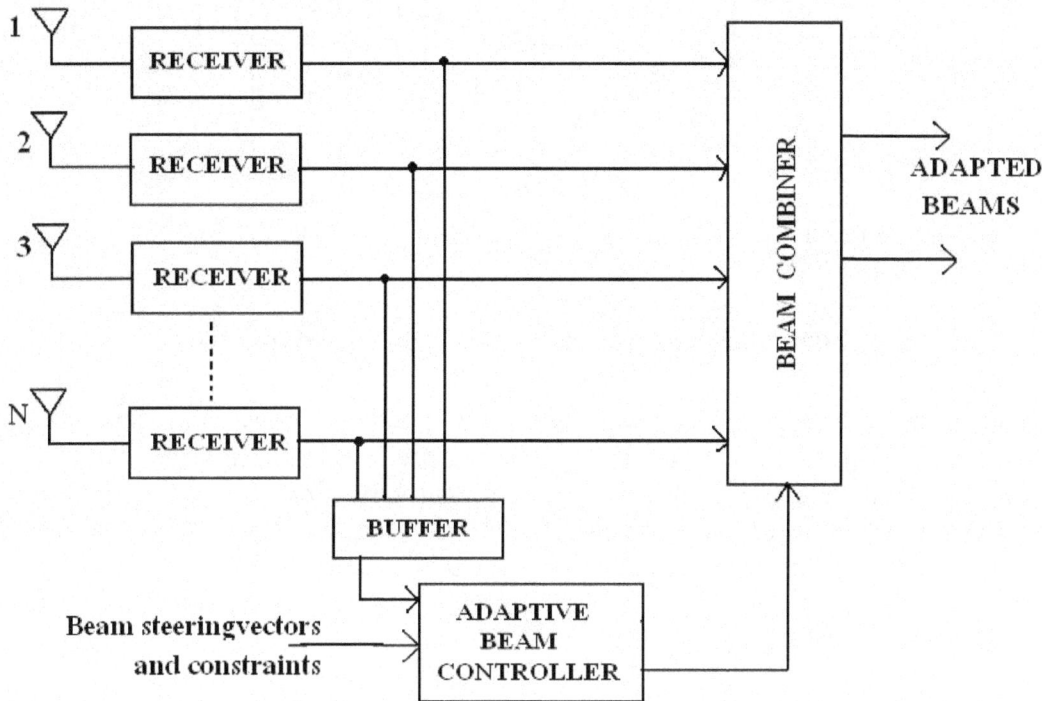

Figure 3.4:- Schematic for Element Space Beam Adaptivity

3.2.2 Beam space beam forming

In contrast to the method of element-space beam forming, where the signals arriving from each of the N elements are weighted and summed to produce the desired output, the beam space technique as shown in figure (3.5) forms multiple fixed beams, using a fixed beam forming networks, which may be spatially orthogonal. The output of each beam is then weighted and the resultant signals are combined to produce the desired output. The signals from the beams, which are not used to supply the desired response, may be used to cancel unknown interference.

Advantage of beam space technique is reduced size of the matrices for adaptive weight computation. In fact, the fixed BFN can improve the performance of the adaptive array processor by providing a certain amount of spatial pre-selection, in which interference arriving from directions away from the desired signal is reduced before applying to the adaptive array processor.

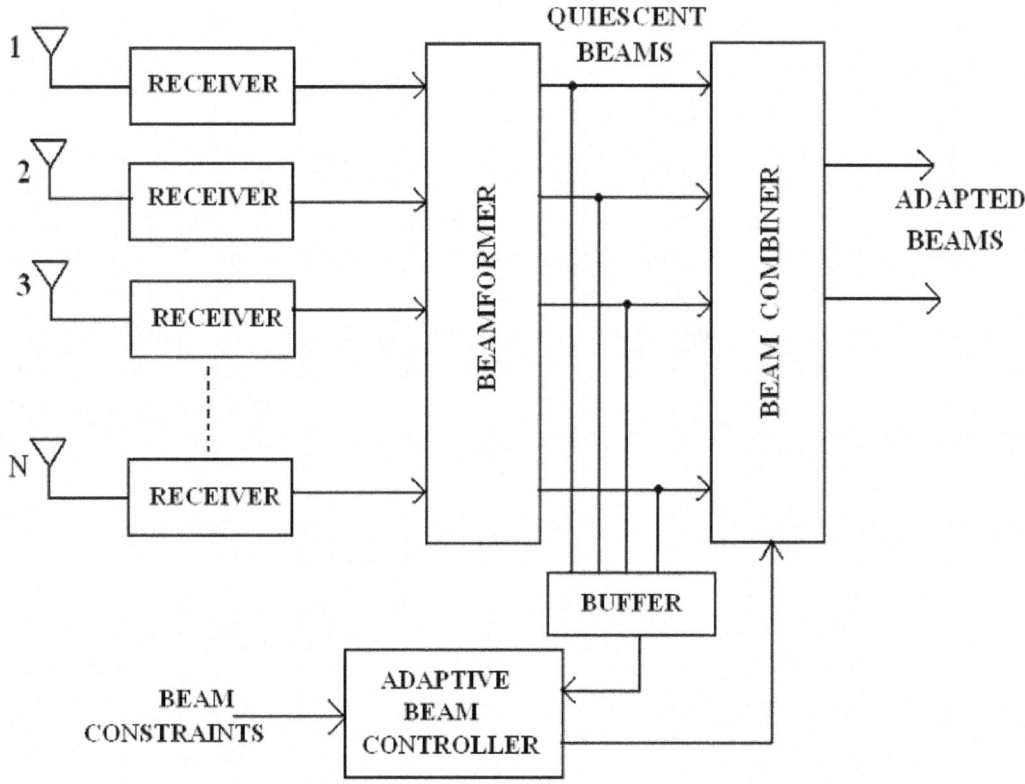

Figure 3.5:-Schematic for Beam Space Adaptivity

This eases dynamic range requirements on the down conversion and sampling systems, and helps the adaptive array initially acquire signals. In particular, since the dynamic range is a major limitation in designing very wide band analogue to digital converters, the use of spatial pre-selection may significantly reduce the cost and difficulty of implementing adaptive array systems.

3.2.2.1 Beam-Space Processing

Beam-space processing is a two stage scheme where the first stage takes the array signals as input and produces a set of multiple outputs, which are then weighted and combined to produce the array output. These multiple outputs may be thought of as the output of multiple beams. The processing done at the first stage is by fixed weighting of the array signals and amounts to produce multiple beams steered in different directions. These weights are normally not adaptive, that is, they are not adjusted during adaptation cycle. The weights applied to different beam outputs to produce the array outputs are optimized to meet a specific optimization criterion and are adjusted during the adaptation cycle. In general, for an -element array, a beam-space processor consists of a main beam steered in the signal direction and a set of not more than (L-1) secondary beams. The weighted output of the secondary beams is subtracted from the main beam. The weights are adjusted to produce an estimate of the interference present in the main beam. The subtraction process then removes this interference. The secondary beams, also known as auxiliary beams, are designed such that they do not contain the desired signal from the look direction to avoid the signal cancellation in the subtraction process. A general structure of such a processor is shown in Fig.3.6.

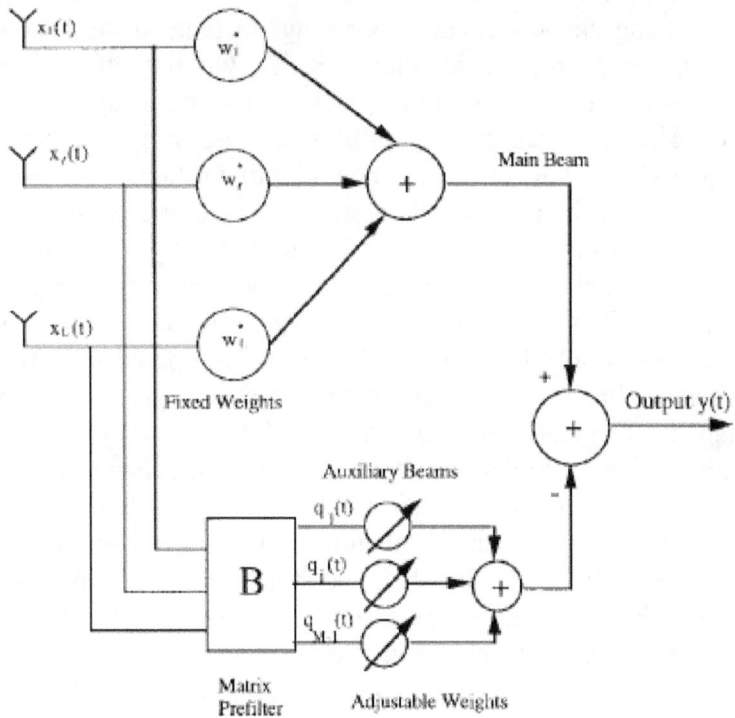

Fig 3.6:- Structure of a general beam space processor

Beam-space processors have been studied under many different names, including Howells–Applebaum array, partitioned processor, partially adaptive arrays adaptive-adaptive arrays and multiple-beam antennas. The pattern of the main beam is normally referred to as the quiescent pattern, and is chosen such that it has a desired shape. For a linear array of equal-spaced elements with equal weighting, the quiescent pattern has the shape of **(sinLx/sinx)** with L being the number of elements in the array, whereas for Chebyshev weighting (the weighting dependent upon the coefficients of the Chebyshev polynomial), the pattern has equal side-lobe levels. The pattern of the main beam may be adjusted by various forms of constraints and pattern synthesis techniques. There are many schemes to generate the outputs of auxiliary beams such that no signal from the look direction is contained in them, that is, the beams have nulls in the look direction. In its simplest form, this can be accomplished by subtracting the array signals from pre-steered adjacent pairs. This relies on the fact that the component of the array signals induced from a source in the look direction is identical after the pre-steering, and this gets canceled in the subtraction process from the adjacent pairs. The process can be generalized to produce (M-1) beams from L element array signals x(t) using a matrix B such that $Q=X^{H(t)}B$ where (M-1) dimensional vector q(t)denotes the outputs of (M-1) beams and the matrix , referred to as the blocking matrix or the matrix pre-filter, has the property that its (M-1) columns are linearly independent and that the sum of the elements of each column

54

equals zero, implying that (M-1) beams are independent and have nulls in the look direction. For an array that is not pre-steered, the matrix needs to Satisfy $\mathbf{S_0^H * B = 0}$, where s_0 is the steering vector associated with the look direction. It is assumed in the above discussion that, M≤L implying that the number of beams is less than or equal to the number of elements in the array. When the number of beams is equal to the number of elements in the array, the processing in the beam space has not reduced the degree of freedom of the array, that is, its null-forming capability has not been reduced. In this sense, these arrays are fully adaptive and have the same capabilities as those of the array using element-space processing. In fact, in the absence of errors, both processing schemes produce identical results. On the other hand, when the number of beams is less than the number of elements, the arrays are referred to as partially adaptive. The null- steering capabilities of these arrays have reduced to that equal to the number of auxiliary beams. When adaptive schemes are used to estimate the weights, the convergence is generally faster for these arrays. The MSE for these arrays, however, is also high compared to that of the fully adaptive arrays. These arrays are useful in situations where the number of interferences is much less than the number of elements. They offer a computational advantage over element-space processing, as one needs only to adjust M-1 weights compared to L weights for the element-space case with M<L Moreover, beam-space processing requires less computation than the element-space case to calculate the weights in general, as it solves an unconstrained optimization compared to the constrained optimization problem solved in the later case. It should be noted that for the element-space processing case, the constraints on the weights are imposed to prevent the signal arriving from the look direction from being distorted and to make the array more robust against errors. For the beam-space case, these are transferred to the main beam, leaving the adjustable weights free from constraints. The beam-space processor considered is a single auxiliary beam processor, referred to as the PIC processor, which is useful for canceling single interference only. The study shows that in the absence of errors, both processors produce identical results, whereas in the presence of look-direction errors, the beam-space processor produces superior performance. The situation arises when the known direction of the signal is different from the actual direction. The weights of the processor are constrained with the knowledge of the look direction. When the actual signal direction is different from the one that is used to constrain weights, the element-space processor cancels this signal as if it was interference close to the look direction. The beam-space processor, on the other hand, is designed to have the main beam steered in the known look direction, and the auxiliary beams are formed to have null in this direction. The response of the main beam does not alter much away from the look direction, and thus the signal level in the main beam is not affected. Similarly when a null of the auxiliary beams is placed in the known look direction then a very small amount of the signal is leaked in the auxiliary beam due to a source very close to the null. Thus, the subtraction process does not affect the signal level in the main beam, yielding a very small signal cancellation in beam-space processing compared to element-space processing. The auxiliary beam-forming techniques other than the use of a blocking matrix (described before) include formation of (M-1) orthogonal beams and formation of beams in the direction of interferences if known. The beams are referred to as orthogonal beams to imply that the weight vectors used to form beams are orthogonal, that is, their dot product is equal to zero. The eigenvectors of R is taken as weights to generate auxiliary beams fall into this category. In situations where the DOA's of interferences are known, the formation of beams pointed in these directions may lead to more

efficient interference cancellation. The auxiliary beam outputs are weighted and summed, and the result is subtracted from the main beam output to cancel the unwanted interference present in the main beam. The weights are adjusted to cancel the maximum possible interference. This is normally done by minimizing the total mean output power after subtraction by solving the unconstrained optimization problem, and leads to maximization of the output SNR in the absence of the desired signal in auxiliary channels. The presence of the signal in these channels causes signal cancellation from the main beam, along with interference cancellation.

3.3 Null steering Beam former

It is used to cancel a plane wave arriving from a known direction and thus produces a null in the response pattern of the plane wave's direction of arrival.
A beam with unity response in the desired direction and nulls in the interference directions may be formed by estimating beam former weights. Assume that S_0 is the steering vector in the direction where unity response is required and that $S_1, ..., S_k$ are k steering vectors associated with k directions where nulls are required. The desired weight vector is the solution of the following simultaneous equation.

$$W^H S_0 = 1 \qquad (3.1)$$

$$W^H S_i = 0, \quad i = 1,...,k \qquad (3.2)$$

Using matrix notation, this becomes

$$W^H A = e_1^T \qquad (3.3)$$

Where A is a matrix with columns being steering vectors associated with all directional sources including the look direction , that is ,

$$A \triangleq [S_0, S_1, ..., S_K]$$

(3.4)

And e_1 is a vector of all zeros except the first element which is one, that is,

$$e_1 = [1,0,...,0]^T \qquad (3.5)$$

For k=L-1, A is a square matrix .Assuming that the inverse of A exists, which requires that all steering vectors are linearly independent, the solution for the weight vector is given by

$$W^H = e_1^T A^{-1} \qquad (3.6)$$

In case the steering vectors are not linearly independent, A is not invertible and its pseudo inverse can be used in its place.

It follows from (3.6) that due to the structure of the vector e_1 the first row of the inverse of matrix 'A' forms the weight vector. Thus the weight selected as the first row of the inverse of matrix 'A' have the desired properties of unity response in the look direction and nulls in the interference direction

When the number of required nulls is less than L-1, A is not a square matrix. A suitable estimate of weights may be produced using

$$W^H = e_1^T A^H \left(AA^H \right)^{-1}$$

(3.7)

Although the beam pattern produced by this beam former has nulls in the interference directions, it is not designed to minimize the uncorrelated noise at the array output. It is possible to achieve this by selecting weights that minimize the mean output power subject to above constraints.

3.4 Desired response and Error

In the adaptation process with performance feedback, the weight vector of the linear combiner is adjusted to cause the output, y_k to agree as closely as possible with the desired response signal. This is accomplished by comparing the output with the desired response to obtain an "error" signal and then adjusting or optimizing the weight vector to minimize this signal. In most practical instances the adaptive process is oriented toward minimizing the mean-square value, on average power of the error signal.

The method of deriving the error signal by means of the desired response input is shown in the figure (3.7). The output signal y_k is simply subtracted from the desired signal d_k to produce the error signal ε_k .

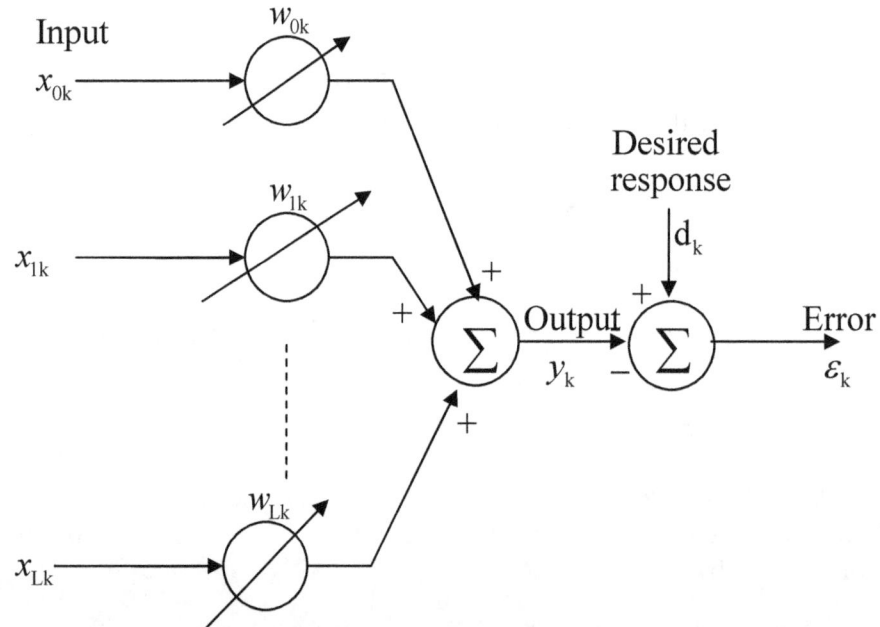

Figure 3.7:-Multiple input adaptive linear combiner with desired Response and error

The input signal vector is represented as

$$X_k = \begin{bmatrix} x_{0k} & x_{1k} & \cdots & x_{Lk} \end{bmatrix}^T$$

(3.8)

The subscript k is used as a time index. Therefore input and output relation is

$$y_k = \sum_{l=0}^{L} w_{lk} x_{lk} \tag{3.9}$$

The weight vector is given by

$$W_k = \begin{bmatrix} w_{0k} & w_{1k} & \cdots & w_{Lk} \end{bmatrix}^T \tag{3.10}$$

Therefore using vector notation

$$y_k = X_k^T W_k = W_k^T X_k \tag{3.11}$$

The source of the desired response signal d_k depends on the application of the adaptive combiner. For the present we will assume the availability of such a signal. As shown in the figure (3.7), the error signal with time index 'k' is

$$\varepsilon_k = d_k - y_k \tag{3.12}$$

Substituting (3.11) into this expression yields

$$\varepsilon_k = d_k - X_k^T W = d_k - W^T X_k \tag{3.13}$$

here we have dropped the subscript 'k' from the weight vector **W** for convenience, because in this discussion we do not wish to adjusts the weights, we now square (3.13) to obtain the instantaneous squared error,

$$\varepsilon_k^2 = d_k^2 + W^T X_k X_k^T W - 2d_k X_k^T W \tag{3.14}$$

We assume that, ε_k, d_k and X_k are statistically stationary and take the expected value of (3.14) over k

$$E[\varepsilon_k^2] = E[d_k^2] + W^T E[X_k X_k^T] W - 2E[d_k X_k^T] W \tag{3.15}$$

Note that, the expected value of any sum of expected values, but that the expected value of a product is the product of expected values when the variables are statistically independent. Signals x_k and d_k are not generally independent.

The mean-square-error function can be more conveniently expressed as follows. Let **R** be defined as the square matrix.

$$R = E[X_k X_k^T] = E \begin{bmatrix} x_{0k}^2 & x_{0k}x_{1k} & x_{0k}x_{2k} & \cdots\cdots & x_{0k}x_{Lk} \\ x_{1k}x_{0k} & x_{1k}^2 & x_{1k}x_{2k} & \cdots\cdots & x_{1k}x_{Lk} \\ \vdots & \vdots & \vdots & \cdots\cdots & \vdots \\ x_{Lk}x_{0k} & x_{Lk}x_{1k} & x_{Lk}x_{2k} & \cdots\cdots & x_{Lk}^2 \end{bmatrix} \tag{3.16}$$

This matrix is designated as "input correlation matrix." The main diagonal terms are the mean squares of the input components and the cross terms are the cross correlations among the input components Let **P** be similarly defined as the column vector

$$P = E[d_k X_k] = E[d_k x_{0k} \quad d_k x_{1k} \quad \cdots\cdots \quad d_k x_{Lk}]^T \tag{3.17}$$

This vector is the set of cross correlations between the desired response and the input components. The elements of both **R** and **P** are all constant second order statistics, when x_k and d_k are stationary. Note that, the multiple input form of x_k was used in (3.16) and (3.17) but the single input form could just as easily have been used.

We now let the mean square error in (3.15) be designated by ε and express it in terms of (3.16) and (3.17) as

$$MSE \triangleq \xi = E[\varepsilon_k^2] = E[d_k^2] + W^T R W - 2P^T W \tag{3.18}$$

It is clear from this expression that the mean square error ε is precisely a quadratic function of the components of the weight vector W when the input components and desired response input are stationary stochastic variables. That is when (3.18) is expanded, the elements of W will appear in first and second degree only.

Gradient and minimum mean square error

Many useful adaptive processes that cause the weight vector to seek the minimum of the performance surface do so by gradient methods [8]. The gradient of the mean square error performance surface designated $\nabla(\xi)$ or simply ∇, can be obtained by differentiating (3.18) to obtain the column vector.

$$\nabla \triangleq \frac{\partial \xi}{\partial W} = \begin{bmatrix} \frac{\partial \xi}{\partial w_0} & \frac{\partial \xi}{\partial w_1} & \cdots\cdots & \frac{\partial \xi}{\partial w_L} \end{bmatrix}^T \tag{3.19}$$

$$= 2RW - 2P \tag{3.20}$$

Where **R** and **P** are given by (3.16) and (3.17) respectively. This expression is obtained by expanding (3.18) and differentiating with respect to each component of the weight vector. Differentiation of the term $W^T R W$ can be treated as differentiation of the product-$(W^T)(RW)$

To obtain the minimum mean-square error the weight vector W is set at its optimal value W^*, where the gradient is zero:

$$\nabla = 0 = 2RW^* - 2P \tag{3.21}$$

Assuming that **R** is non-singular, the optimal weight vector W^*, sometimes called the Weiner weight vector, is found from (3.21) to be

$$W^* = R^{-1}P \tag{3.22}$$

This equation is an expression of the Weiner and -Hopf equation in matrix form. The minimum mean-square error is now obtained by substituting W^* from (3.22) for W in (3.18):

$$\xi_{\min} = E\left[d_k^2\right] + W^{*T}RW^* - 2P^TW^*$$

$$= E\left[d_k^2\right] + \left[R^{-1}P\right]^T RR^{-1}P - 2P^T R^{-1}P \tag{3.23}$$

3.5 The LMS Algorithm:

One way of finding the optimum set of weight values is to solve equation (3.22). This solution is generally straightforward presents serious computational problems when the number of weights n is large and when data rates are high. In addition to the necessity of inverting $n \times n$ matrix, this method may require as many as $n(n+1)/2$ autocorrelation and cross correlation measurements to obtain the elements of matrix R. Furthermore, this process generally needs to be continually repeated in most practical situations where the input signal statistics change slowly. No perfect solution of (3.22) is possible in practice because of the fact that an infinite statistical sample would be required to estimate perfectly the elements of the correlation matrices.

The LMS algorithm is one method for finding approximate solution to (3.22). The accuracy is limited by statistical sample size, since the algorithm finds the weight values based on finite time measurements of input data signals. This method does not require explicit measurements of correlation functions or matrix inversion. It is based on gradient search techniques applied to mean square error function. The LMS algorithm does not require squaring, averaging or differentiation in order to make use of gradients of mean-square-error functions.

The LMS algorithm is based on the method of steepest descent. Changes in the weight vector are made along the direction of the estimated gradient vector. Accordingly,

$$W(j+1) = W(j) + k_s \tilde{\nabla}(j) \tag{3.24}$$

Where,

$W(j)$=weight vector before adaptation.

$W(j+1)$ = weight vector after adaptation.

k_s = scalar constant controlling rate of convergence and stability ($k_s < 0$)

$\tilde{\nabla}(j)$ = estimated gradient vector of $\bar{\varepsilon}^2$ with respect to W.

One method for obtaining the estimated gradient of the mean-square-error function is to take the gradient of a single time sample of the squared error.

$$\tilde{\nabla}(j) = \nabla[\varepsilon^2(j)] = 2\varepsilon(j)\nabla[\varepsilon(j)]$$

From (3.5)

$$\nabla[\varepsilon(j)] = \nabla[d(j) - W^T(j)X(j)]$$
$$= -X(j)$$

Thus

$$\tilde{\nabla}(j) = -2\varepsilon(j)X(j) \qquad (3.25)$$

The gradient estimate of (3.25) is unbiased, as will be shown by the following argument. For a given weight vector $W(j)$, expected value of the gradient estimate is

$$E[\tilde{\nabla}(j)] = -2E[\{d(j) - W^T(j)X(j)\}X(j)]$$

$$= -2E[P - W^T(j)R] \qquad (3.26)$$

Comparing (3.20) and (3.26), we see that

$$E[\tilde{\nabla}(j)] = \nabla E[\varepsilon^2]$$

and therefore, for a given weight vector, the expected value of the estimate equals the true value. Using the gradient estimation formula given in (3.25), the weight iteration rule (3.24) becomes

$$W(j+1) = W(j) - 2k_s\varepsilon(j)X(j) \qquad (3.27)$$

and the next weight vector is obtained by adding to the present weight vector the input vector scaled by the value of the error.

The LMS algorithm is given by (3.27). It is directly usable as a weight adaptation formula for digital systems.

3.6 Non blind Adaptive Beam-former

This type of adaptive beam-former is based on a "pilot signal." The pilot signal adaptive beam former forms the beam toward a specified "look direction" and uses adaptivity to support this beam while simultaneously forming notches to null interference arriving outside the look direction. The nulling process is determined by direction of arrival and power level. While the system is adapting, an injected pilot signal simulates a received signal from a look direction chosen by the system operator. The same pilot signal is used as the desired response for the adaptive processor attached to the antenna array elements. Using the pilot signal, the adaptive beam former is trained so that its directivity pattern has main lobe in the specified look direction as well as notches corresponding with incident interference signals whose directions of the arrival differ from the look direction. The array thus adapts to form the main lobe with its direction and bandwidth determined by the pilot signal which arrives outside of the main lobe. In short, Non blind adaptive algorithms require pilot signal and information about the DOA of desired signal. The LMS algorithm described above is class of non blind adaptive beam-former.

3.7 Blind Adaptive Beam-former

In this type of algorithms training or pilot signal is not required. This class of algorithm usually requires DOA (Direction Of Arrival) of desired signal only. Information regarding the DOAs of interferers and their power levels is not required. Depending upon the strength and DOAs of interferers, blind adaptive beam-former automatically forms nulls in the direction of interferers. The Direct Data Domain Least Square algorithms based on the single snap shot of data, such as Forward method or Backward method or Forward-Backward method algorithms, fall under the category of blind adaptive beam-former.

3.8 SMI Algorithm

This algorithm estimates the array weights by replacing R with its estimate. An unbiased estimate of R using N samples x(n), n=1,2,.... N-1 of the array signals may be obtained using a simple averaging scheme.

$$R(n) = \frac{1}{N} \sum_{n=0}^{N-1} x(n)x^H(n)$$

Where $R(n)$ denotes the estimate at the th instant of time $x(n)$ and denotes the array signal sample, also known as the array snapshot, at the nth instant of time, with t replaced by nT and the sampling time T omitted for the ease of notation. The estimate of R may be updated when the new samples arrive using the equation.

$$R(n+1) = \frac{nR(n) + x(n+1)x^H(n+1)}{n+1}$$

And a new estimate of the weights w(n+1) at time instant (n+1) may be made. The expression of the optimal weights requires the inverse of R, and this process of estimating R and then its inverse may be combined to update the inverse of R from array signal samples using the Matrix Inversion Lemma as follows:

$$R^{-1}(n) = R^{-1}(n-1) - \frac{R^{-1}(n-1)x(n)x^H(n)R^{-1}(n-1)}{1 + x^H(n)R^{-1}(n-1)x(n)}$$

$$\text{With } R^{-1}(0) = \frac{1}{\varepsilon_0}I, ...\varepsilon_0 > 0$$

This scheme of estimating weights using the inverse update is referred to as the RLS algorithm. It should be noted that as the number of samples grows, the matrix update approaches its true value, and thus the estimated weights approach the optimal weights, that is, as $n \to \infty, R(n) \to R,$ and $w(n) \to \hat{w}$ or w_{MSE}

Procedures for estimating array weights with efficient computation using SMI show how it performs as a function of the number of snapshots. Application of SMI to estimate the weights of an array to operate in mobile communications particularly for GSM signals. (using a variable reference signal as available during the symbol interval of the TDMA system)

3.9 RLS Algorithm

The convergence of the LMS algorithm depends upon the Eigen values of R. In an environment yielding R with a large Eigen value spread, the algorithm converges with slow speed. This problem is solved in an RLS algorithm by replacing the gradient step size with a gain Matrix $R^{-1}(n)$ at the n th iteration, producing the weight update equation: $w(n) = w(n+1) - R^{-1}(n)x(n)\varepsilon^*(w(n-1))$

Where R(n) is given by $R(n) = \delta_0 R(n-1) + x(n)x^H(n) = \sum_{k=0}^{n} \delta_0^{n-k} x(k)x^H(k)$

Where, δ_0 a real scalar smaller than but close to one, is used for exponential weighting of the past data and is referred to as the forgetting factor, as the update equation tends to deemphasize the old samples. The quantity $(1/1- \delta_0)$ is normally referred to as the memory of the algorithm. Thus, for $\delta_0 = .99$, the memory of the algorithm is close to 100
samples. The RLS algorithm updates the required inverse of R(n) using the previous inverse and the present sample as:

$$R^{-1}(n) = \frac{1}{\delta_0}[R^{-1}(n-1) - \frac{R^{-1}(n-1)x(n)x^H(n)R^{-1}(n-1)}{\delta_0 + x^H(n)R^{-1}(n-1)x(n)}$$

The matrix is initialized as $R^{-1}(0) = \frac{1}{\varepsilon_0}I, ...\varepsilon_0 \rangle 0$

The RLS algorithm minimizes the cumulative square error.

$$J(n) = \sum_{k=0}^{n} \delta_0^{n-k} |\varepsilon(k)|^2$$

And its convergence is independent of the Eigen value distribution of the correlation matrix. The algorithm presented here is the exact RLS algorithm.
A comparison of the convergence speed of the LMS, the RLS, and some other gradient-based algorithms using quantized or clipped data indicates that RLS is the most efficient and LMS is the slowest. A computer-simulation study of the RLS, LMS, and SMI algorithms in a mobile communications situation suggests that the former outperforms the latter two in flat-fading channels. An application of the RLS algorithm for the reverse link of a cellular communication using the CDMA system increase channel capacity with an adaptive array.

3.10 CMA

CMA is a gradient-based algorithm that works on the premise that the existence of an interference causes fluctuation in the amplitude of the array output, which otherwise has a constant modulus. It updates the weights by minimizing the cost function:

$$J(n) = \frac{1}{2}E[(|y(n)|^2 - y_0^2)^2]$$

Using the following equation: $w(n+1) = w(n) - \mu g(w(n))$

Where $y(n) = w^H(n)x(n+1)$. The array output is after the nth iteration, y_0 is the desired amplitude in the absence of interference, and $g(w(n))$ denotes an estimate of the gradient of the cost function. Similar to the LMS algorithm discussed previously, it uses an estimate of the gradient by replacing the true gradient with an instant value given by

$g(w(n)) = 2\varepsilon(n)x(n+1)$ Where $\varepsilon(n) \triangleq (|y(n)|^2 - y_0^2)y(n)$. The weight update equation for this case becomes:

$$w(n+1) = w(n) - 2\mu\varepsilon(n)x(n+1)$$

In appearance, this is similar to the LMS algorithm with a reference signal where

$$\varepsilon(n) \triangleq d(n) - y(n)$$

Its application to a digital land-mobile radio communications system using TDMA is studied to compensate for selective fading. CMA for beam-space array signal processing, indicate that the beam-space CMA is able to cancel interferences arriving from directions other than the look direction. CMA is useful for eliminating correlated arrivals and is effective for constant modulated envelope signals such as GMSK and QPSK, which are used in digital communications. The algorithm, however, is not appropriate for the CDMA system because of the required power control. Differential CMA has inferior convergence characteristics compared to CMA but may be improved using DOA information to make it operative in beam space.

3.8 Simulation Results

3.8.1 Simulation of null steering beam former with four elements antenna array.

Consider four elements linear array consisting of omnidirectional antenna elements spaced $\lambda \backslash 2$ distance apart. With four elements in the antenna array we can specify four positions in the array pattern. These may be unity response in the desired direction and three nulls in the directions where interfering noise sources are likely to be present. It is also possible to have unity response in the 2 or 3 directions and nulls in the 2 or 1 direction respectively. In total we can specify four positions in the radiation pattern.

Array factor of four element linear array with centre as a reference point is given by

$$A.F = W_1 * \exp(-j1.5 * si) + W_2 * \exp(-j0.5 * si) + W_3 * \exp(j0.5 * si) + W_4 * \exp(j1.5 * si)$$

Where,
$$si = k * d * \sin(\theta)$$
$$k = 2\pi / \lambda$$
d = inter element spacing

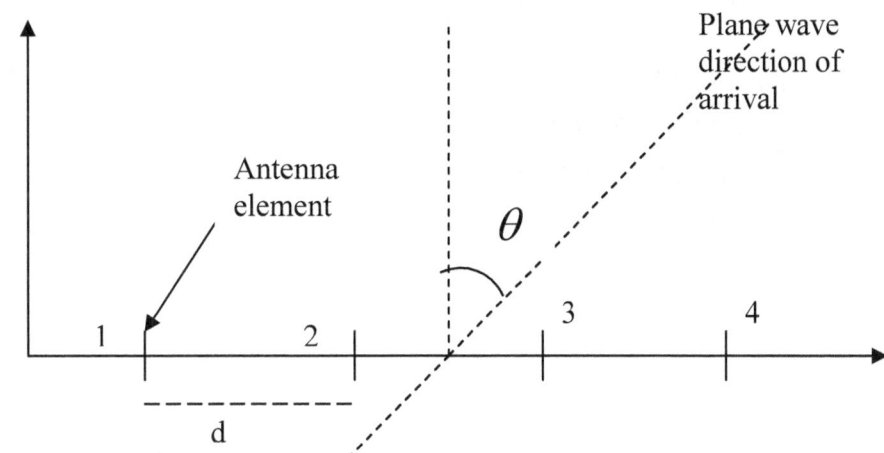

Figure 3.8:- Four elements equally spaced linear array

Suppose we require

A.F = 1   at $\theta = \pi / 4$

64

$$= 0 \qquad \text{at} \quad \theta = \pi/2, \quad \pi/6, \quad 0$$

Assume $d = \lambda/2$

Then steering vector is

$S_i=[\exp(-j1.5kd\sin(\theta_i)) \quad , \quad \exp(-j0.5kd\sin(\theta_i)) \quad , \quad \exp(j0.5kd\sin(\theta_i)) \quad ,$
$\exp(j1.5kd\sin(\theta_i)) \,]$

At $\theta = \pi/4$ (look direction of antenna array)
The steering vector associated with look direction is

$S_0=[\exp(-j1.5kd\sin(\pi/4)) \quad , \quad \exp(-j0.5kd\sin(\pi/4)) \quad , \quad \exp(j0.5kd\sin(\pi/4)) \quad ,$
$\exp(j1.5kd\sin(\pi/4)) \,]^T$

$= [-0.9819-0.1894i, \ 0.444+0.896i, \ 0.444-0.896i, \ -0.9819+0.1894i]^T$

Similarly steering vectors associated with null directions at $\theta = \pi/2, \quad \pi/6, \quad 0$
$S_1 = [0.0-1i, \ 0.0+1i, \ 0.0-1i, \ 0.0+1i]^T$

$S_2 = [-0.707+0.707i, \ 0.707+0.707i, \ 0.707-0.707i, \ -0.707-0.707i]^T$

$S_3 = [1+0.0i, \ 1+0.0i, \ 1+0.0i, \ 1+0.0i]^T$

$A= [S_0, S_1, S_2, S_3]$

$e_1= [1, 0, 0, 0]$

Weight vector W is given by

$W^H = e_1^T \ \text{inv} \ (A)$

$$W = \begin{bmatrix} w_1 \\ w_2 \\ w_3 \\ w_4 \end{bmatrix} = \begin{bmatrix} -0.695 - 0.695i \\ 0.695 - 0.695i \\ 0.695 + 0.695i \\ -0.695 + 0.695i \end{bmatrix}$$

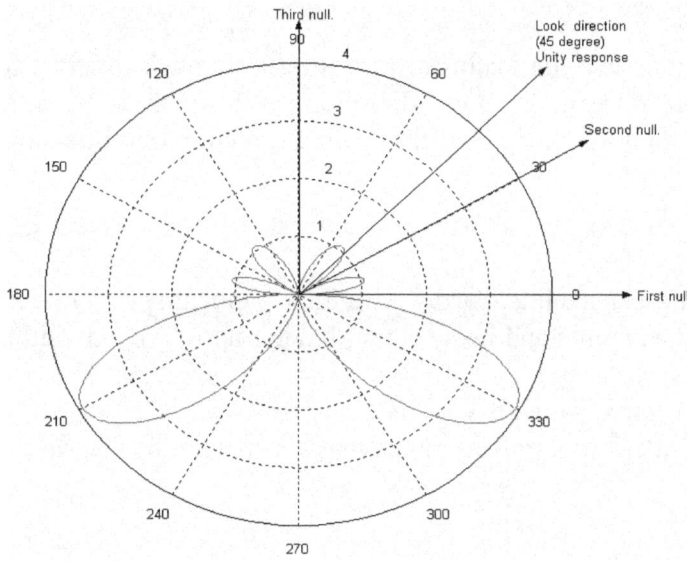

Figure 3.9:- Simulation result of simple null steering beam former using four elements linear array.

Analysis of simulation result of null steering beam-former

1) Maximum value in the array factor response is function of weights as well as direction.

2) With four elements adaptive array we have control over only four directions, where we can obtain desired response, but in the directions other than these four directions, response of the array factor may be greater than response of the look direction.

3) We can get exact value of response in the look direction that we expected (here we expected unity response). We also expect, in the directions other than look directions response should be very less than response in the look direction, but we didn't get that.

4) The maximum value of array factor response in the direction other than look direction is a function of both weights calculated and an angle θ.

Limitations of simple null steering beam-former

From the simulation result in figure 3.9

1) We get unity response in the look direction and nulls in the desired directions, but there   may be greater than unity response in the directions other than look and null directions, which is undesirable.

In the transmission mode, such type of simple null steering beam-former wastes energy in the directions other than look direction.

2) It requires knowledge of the directions of interfering sources.

3) The weights estimated by this scheme do not maximize the output SNR.

4) If small noise component presents in the direction where array response is greater than unity (e.g. at $\theta = \pi/4$), then this noise component gets a gain over the look direction and will reduce the output SNR of adaptive array.

66

Conclusion

Using simple null steering beam-former, it is very much difficult to get maximum array factor response in the look direction, nulls in the interfering noise source directions and minimum response in the other directions simultaneously.

3.8.2 Simulation results of adaptive array using LMS algorithm

Assumptions:-

1) Assume four elements linear array with inter-element spacing d=$\lambda \setminus 2$
2) Assume desired signal and noise signal having narrow bandwidth and at the same frequency.
3) Direction of desired signal is known.
4) Direction of interfering noise sources may or may not be known.

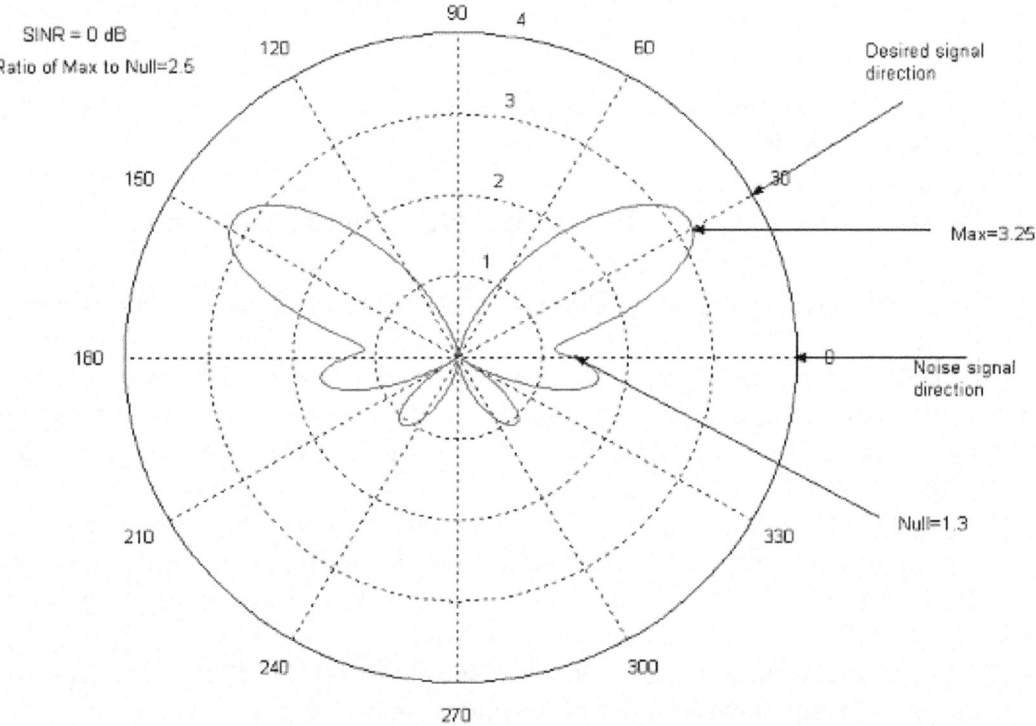

Figure 3.10:- Simulation result of four elements linear adaptive array using LMS algorithm (when SINR= 0 dB)

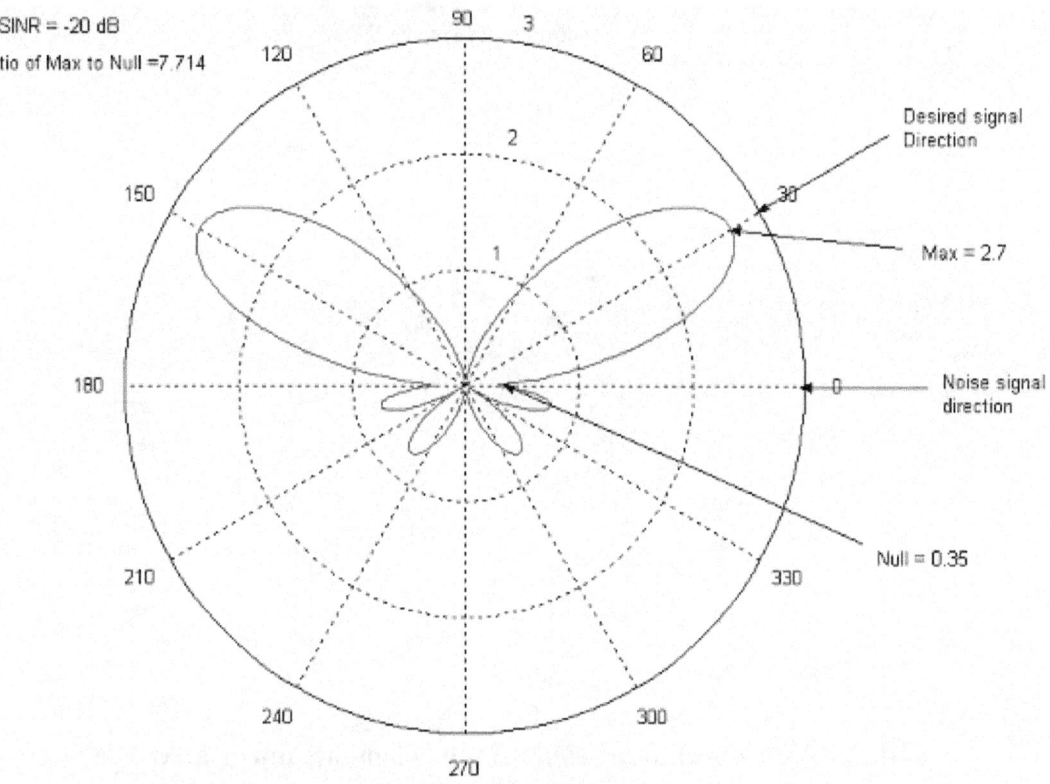

Figure 3.11:- Simulation result of four elements linear adaptive array using LMS Algorithm (when SINR= -20 dB)

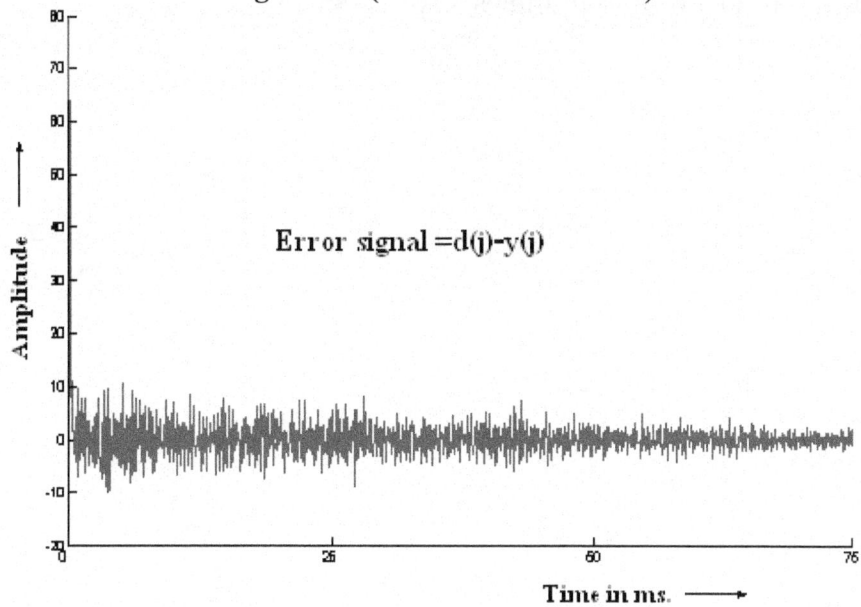

Figure 3.12: Error signal generated during simulation of LMS algorithm

Conclusions:

1) Deep null in the direction of interfering signal is formed when noise amplitude is high.
2) After finite number of iterations LMS algorithm converges to optimal solution.
3) Null formation in the direction of interfering noise source becomes difficult when it is closer to the look direction.

68

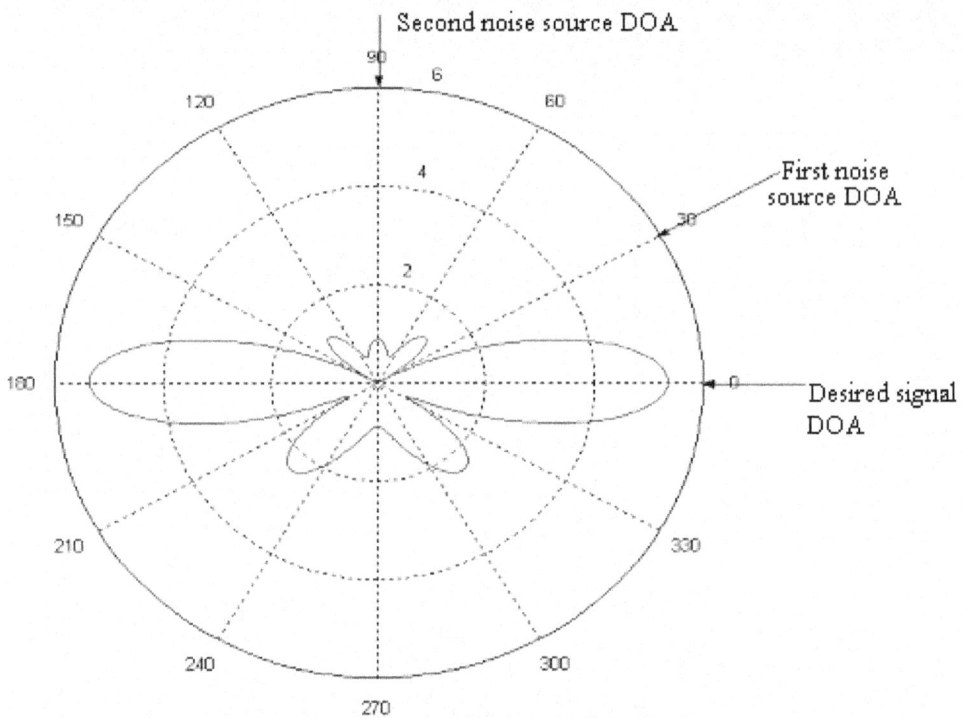

Figure 3.13:- Simulation result of four elements linear adaptive array using LMS algorithm

3.8.3 Simulation of 6 by 6 planer array

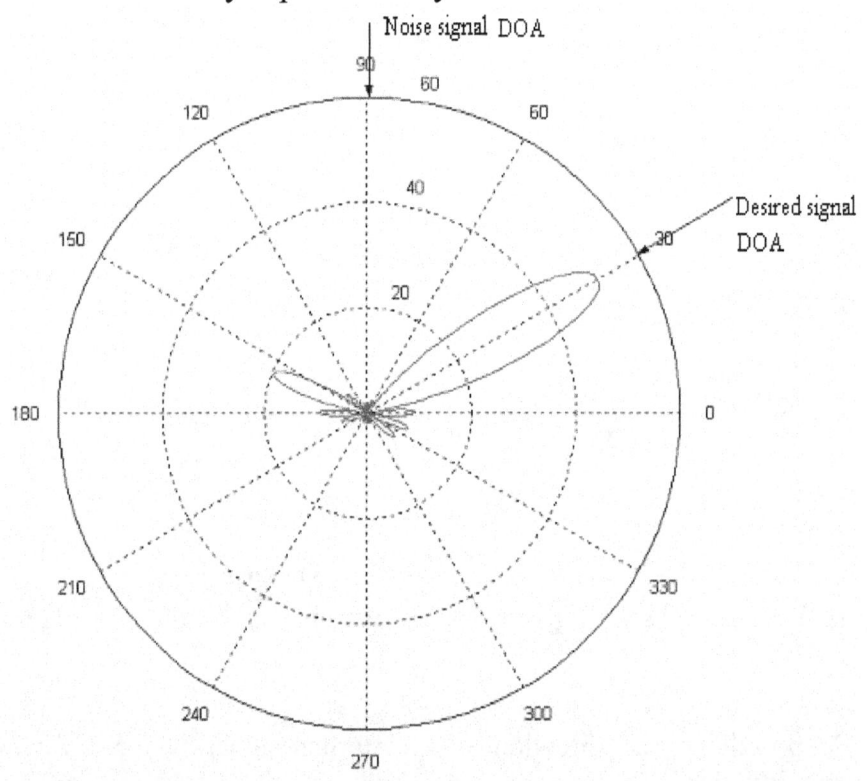

Figure 3.14:- Simulation result of 6 by 6 elements planer adaptive array using LMS algorithm

CHAPTER 4

Direct Data Domain Least Squares approaches to Adaptive Processing based on Single Snapshots of Data.

An adaptive beam-former is a device that is able to separate signals which have the same frequency content but are separated in the spatial domain. This provides a means for separating a desired signal from interfering signals. An adaptive beam-former is able to optimize the array pattern automatically by adjusting the weights associated with each antenna element until a prescribed objective function is satisfied.

4.1 Introduction

In this chapter a direct data domain least squares (D^3LS) approach to adaptive processing using a single snapshot of a data is presented. In contrast to the conventional adaptive techniques, where at the first step one needs to form the covariance matrix of the data, and then invert it, in the present approaches the signal of interest (SOI) is obtained directly from the solution of a matrix equation. Use of the conjugate gradient in solving this matrix equation makes this procedure highly suitable to real-time implementation of these algorithms, as the computational time is considerably less than that of the other conventional adaptive techniques. Here a least squares approach is applied directly to the data on a snapshot-by-snapshot basis and hence is computationally quite efficient. A snapshot in this specific case is defined as the phasor voltages measured at the feed points of all the antenna elements in the array at a particular instance of a time. Nonstationarity in the data then has little effect for these classes of direct data domain methods as no assumption is made about the statistics of the environment. Yet the environment is modeled in a realistic fashion. Even though we are processing the data on a snapshot-by snapshot basis, one can still extract N coherent sources using an antenna array consisting of approximately 1.5 N elements which is the minimum possible under any circumstances. However the direct data domain methods is result in a slightly reduced number of degrees of freedom as opposed to a conventional statistical analysis for noncoherent interferers, where one needs to process a block of data snapshots to form the covariance matrix. In addition it is shown how to carry out adaptive processing with a pre-specified main lobe width of the adapted beam pattern thereby preventing signal cancellation in the direct data domain algorithms.
In D^3LS algorithms direction of arrival of the signal of interest (SOI) should be known a priori. In addition to the SOI contributing to the received voltages at each antenna element, we also have contributions due to interferers, and thermal noise. The incoming interferers may be coherent with the SOI. Consider a uniformly spaced linear array consisting of N+1 isotropic omnidirectional point radiators as shown in figure (4.1).

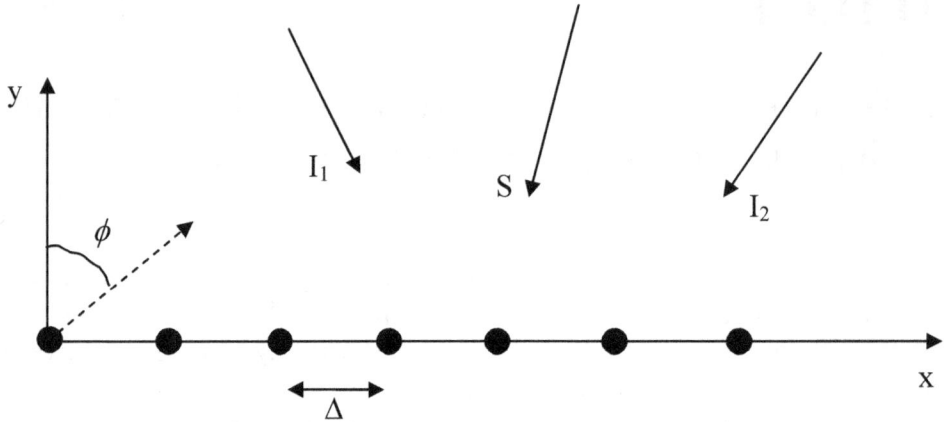

Figure 4.1:- Linear uniform array.

The phasor voltage Xn induced at the nth antenna element at a particular instance of time will then be given by

$$X_n = S\,e^{\frac{j2\pi n\Delta\sin\phi_S}{\lambda}} + \sum_{p=1}^{P} I_p e^{\frac{j2\pi n\Delta\sin\phi_p}{\lambda}} + \xi_n \tag{4.1}$$

where,

S = complex amplitude of the SOI (to be determined)

ϕ_S = direction of arrival of the SOI (assumed to be known)

Δ = spacing between each of the antenna elements

λ = wavelength of transmission (here it is assumed that we are dealing with narrowband signals)

P = total number of interferers

A_p = complex amplitude of pth undesired interferer

ϕ_p = direction of arrival of pth interferer

ξ_n = thermal noise induced at the nth antenna element

For the array shown in Figure 4.1, the measured voltages X_n for n=0, 1, 2,......,N, at the antenna elements are assumed to be known along with θ_s, the direction of arrival of the SOI. The goal is to estimate the complex amplitude S for the SOI. Here we define a single snapshot by the voltages X_n, measured at the nth element at a certain instant of time t_m. It is understood that all the SOI, interferers and thermal noise vary as a function of time.

4.2 Direct Data Domain in Least Squares Procedures

Consider the linear array as shown in figure (4.1). Here we consider that we have a single snapshot of the voltages measured at the feed point of the antenna elements (i.e., at a time t = t$_m$, we have the voltages X$_n$, for n = 0, 1,.........,N measured at the feed points of all N+1 antenna elements). Our goal is to estimate s givenφ_s . We also know that in order to obtain the SOI, the number of coherent interferers must be $\leq N/2$ in the absence of noise. It is important to point out that in this

71

procedure we do not make any distinction between coherent or noncoherent interferers.

4.2.1 Forward Method

Under the assumption that number of interferers $P \leq \dfrac{N}{2}$, where N+1 is number of antenna elements in the array and P is number of interferers. As we have to calculate unknown weights from single snap-shot of data then it is not possible to calculate all N+1 unknown weights , but we can calculate L+1 weights $(L+1 < N+1)$ where L+1 represents sub-array of L+1 elements.

Therefore one can form matrix pencil $[X] - \alpha[S]$ of dimension L+1. Here α is the estimate of the complex amplitude for the unknown SOI to be solved. Where,

$$[X] = \begin{bmatrix} X_0 & X_1 & \cdots\cdots & X_L \\ X_1 & X_2 & \cdots\cdots & X_{L+1} \\ \vdots & \vdots & \cdots\cdots & \vdots \\ X_L & X_{L+1} & \cdots\cdots & X_N \end{bmatrix}_{(L+1)\times(L+1)}$$

(4.2)

The matrix defined by equation (4.2) has a form of Hankel matrix. The peculiarity of Hankel matrix is that its anti diagonal elements are same.

$$[S] = \begin{bmatrix} S_0 & S_1 & \cdots\cdots & S_L \\ S_1 & S_2 & \cdots\cdots & S_{L+1} \\ \vdots & \vdots & \cdots\cdots & \vdots \\ S_L & S_{L+1} & \cdots\cdots & S_N \end{bmatrix}_{(L+1)\times(L+1)}$$

(4.3)

In actual practice carrier frequency of transmitter is known in most of the cases. As shown by equation (4.1) at the output of the antenna element total voltage is available. Total voltage contains contribution from signal of interest (SOI), interferers and thermal noise. Therefore if α is estimated then the variables in the (4.2) and (4.3) are defined in such a way that the difference $X_n - \alpha S_n \, (n = 0...N)$ at each element represents the contribution due to signal multipaths, interferers and thermal noise (i.e. all undesired components of the signals expect the signal of interest, which is α). Since it is assumed that S_n is the voltage induced at the nth element due to a signal arriving from the same direction as the SOI but whose amplitude is unity
Therefore,

$$S_n = e^{\dfrac{j2\pi n \Delta \sin \phi_s}{\lambda}}$$

In an adaptive processing methodology, the column vector of weights $[W]$ is chosen in such a way that the contribution from the jammers or interferers and thermal noise are minimized to enhance the output signal to interference plus noise ratio. Hence if we define the matrix,

$$[U] = \{ [X] - \alpha[S] \}$$ (4.4)

We obtain the following generalized eigenvalue problem:

$$[U]_{(L+1,L+1)}[W]_{L+1} = \{[X] - \alpha[S]\}_{(L+1,L+1)}[W]_{L+1} = 0$$ (4.5)

Now here problem comes, weight vector as well as α are unknown. So we will apply one intelligent trick to get rid of α for the calculation of unknown weight vector W. Please remember that α is our objective, so we estimate α after calculation of unknown weight vector. At this stage Forward Method of D^3LS procedure starts.

First of all let us see how to calculate unknown weights vector W without the knowledge of α estimate. Note that the (1, 1) and (1, 2) elements of the interference plus noise matrix $[U]$ as defined in equation (4.5) are given by

$$U(1,1) = X_0 - \alpha S_0$$ (4.6)

$$U(1,2) = X_1 - \alpha S_1$$ (4.7)

where X_0 and X_1 are the voltages received at antenna elements 0 and 1 due to the signal, interferer and thermal noise whereas S_0 and S_1 are the values of the SOI only at those elements due to a signal of unit strength. Define

$$Z = \exp[\frac{j2\pi d}{\lambda}\sin\phi_s]$$ (4.8)

Then $U(1,1) - Z^{-1}U(1,2)$ contains no components of the SOI, as

$$S_0 = \exp\left[\frac{j2\pi(n=0)d}{\lambda}\sin\phi_s\right] \quad \text{with n} = 0$$ (4.9)

and

$$S_1 = \exp\left[\frac{j2\pi(n=1)d}{\lambda}\sin\phi_s\right] \quad \text{with n} = 1$$ (4.10)

The same is true for $U(1,2) - Z^{-1}U(1,3)$ and in general, for $U(i,j) - Z^{-1}U(i,j+1)$ for $i = 1,..........L+1$ and $j = 1,..........L$. Here we have L=N/2. Therefore one can form a reduced rank matrix $[T]_{L\times(L+1)}$ generated from $[U]$ such that

$$[T] = \begin{bmatrix} X_0 - Z^{-1}X_1 & X_1 - Z^{-1}X_2 & \cdots\cdots & X_L - Z^{-1}X_{L+1} \\ X_1 - Z^{-1}X_2 & X_2 - Z^{-1}X_3 & \cdots\cdots & X_{L+1} - Z^{-1}X_{L+2} \\ \vdots & \vdots & \cdots\cdots & \vdots \\ X_{N-L-1} - Z^{-1}X_{N-L} & X_{N-L} - Z^{-1}X_{N-L+1} & \cdots\cdots & X_{N-1} - Z^{-1}X_N \end{bmatrix}_{(L)\times(L+1)}$$

$$\times \begin{bmatrix} W_0 \\ W_1 \\ \vdots \\ W_L \end{bmatrix} = 0$$

(4.11)

In order to restore the signal component in the adaptive processing, we fix the gain of the sub array formed by the L+1 element along the direction ϕ_s. That is we constrain the gain of the sub array formed by L+1 element in the direction of the arrival of SOI. Let us say that the gain of the sub array is C along the direction of ϕ_s. This provides an additional equation resulting in a square matrix.

$$\begin{bmatrix} 1 & Z & \cdots\cdots & Z^L \\ X_0 - Z^{-1}X_1 & X_1 - Z^{-1}X_2 & \cdots\cdots & X_L - Z^{-1}X_{L+1} \\ X_1 - Z^{-1}X_2 & X_2 - Z^{-1}X_3 & \cdots\cdots & X_{L+1} - Z^{-1}X_{L+2} \\ \vdots & \vdots & \cdots\cdots & \vdots \\ X_{N-L-1} - Z^{-1}X_{N-L} & X_{N-L} - Z^{-1}X_{N-L+1} & \cdots\cdots & X_{N-1} - Z^{-1}X_N \end{bmatrix}_{(L+1)\times(L+1)}$$

$$\times \begin{bmatrix} W_0 \\ W_1 \\ \vdots \\ W_L \end{bmatrix}_{(L+1)\times1} = \begin{bmatrix} C \\ 0 \\ \vdots \\ 0 \end{bmatrix}_{(L+1)\times1}$$

(4.12)

or equivalently

$$[A][W] = [Y]$$

(4.13)

Once the weights are solved by using (4.13), the signal component α (estimate of SOI in (4.1)), is estimated by using

$$\alpha = \frac{1}{C} \sum_{i=0}^{L} W_i X_i$$

(4.14)

4.2.2 Backward method

Next we reformulate the problem using the same data to obtain a second independent estimate for the solution. This is achieved by reversing the data sequence and then complex conjugating each term of that sequence.

It is well known in the parametric spectral estimation literature that a sampled sequence, which can be represented by a sum of exponentials with purely imaginary argument, can be used in either the forward or reverse direction, resulting in the same value for the exponent. Therefore, whether we look at the snapshot as a forward sequence as a presented in the previous section or by a reverse conjugate of the same sequence, the final results for W_i must be the same. Hence for these classes of problems, we can observe the data in either the forward or reverse direction. This is equivalent to creating a virtual array of the same size but located along a mirror symmetry line. Therefore, if we now conjugate the data and form that reverse sequence, we get an independent set of equations similar to (4.12) for the solution of the weights [W]. This is represented by

$$
\begin{bmatrix}
1 & Z & \cdots\cdots & Z^L \\
X_N^* - Z^{-1}X_{N-1}^* & X_{N-1}^* - Z^{-1}X_{N-2}^* & \cdots\cdots & X_L^* - Z^{-1}X_{L-1}^* \\
\vdots & \vdots & \cdots\cdots & \vdots \\
X_{L+1}^* - Z^{-1}X_L^* & X_L^* - Z^{-1}X_{L-1}^* & \cdots\cdots & X_1^* - Z^{-1}X_0^*
\end{bmatrix}_{(L+1)\times(L+1)}
$$

$$
\times
\begin{bmatrix}
W_0 \\
W_1 \\
\vdots \\
W_L
\end{bmatrix}_{(L+1)\times 1}
=
\begin{bmatrix}
C' \\
0 \\
\vdots \\
0
\end{bmatrix}_{(L+1)\times 1}
\tag{4.15}
$$

or equivalently, in matrix form as

$$
[B][W] = [Y] \tag{4.16}
$$

The signal strength α can now be determined from

$$
\alpha_v = \left[\frac{Z^{L+v}}{C'} \sum_{i=0}^{L} W_i X_{L-i+v}^* \right]^* \quad \text{for} \quad v = 0,\ 1, \ldots\ldots\ldots L \tag{4.17}
$$

Note that for both the forward and backward methods, we have $L = N / 2$. Hence the degrees of freedom are the same for both the Forward and Backward methods. However, we have two independent solutions for the same adaptive problem. In a real situation, when the solution is unknown, two different estimates for the same solution may provide a level of confidence on the quality of the solution.

4.2.3 Forward-Backward method

Now we combine the forward and backward methods to double the given data and thereby increase the number of weights or degrees of freedom significantly over that of either the forward or backward method alone.

$$
\begin{bmatrix}
1 & Z & \cdots\cdots & Z^{Q} \\
X_0 - Z^{-1}X_1 & X_1 - Z^{-1}X_2 & \cdots\cdots & X_V - Z^{-1}X_{V+1} \\
X_1 - Z^{-1}X_2 & X_2 - Z^{-1}X_3 & \cdots\cdots & X_{V+1} - Z^{-1}X_{V+2} \\
\vdots & \vdots & \cdots\cdots & \vdots \\
X_{N-Q-1} - Z^{-1}X_{N-Q} & X_{N-Q} - Z^{-1}X_{N-Q+1} & \cdots\cdots & X_{N-1} - Z^{-1}X_N \\
X_N^* - Z^{-1}X_{N-1}^* & X_{N-1}^* - Z^{-1}X_{N-2}^* & \cdots\cdots & X_{N-Q}^* - Z^{-1}X_{N-Q-1}^* \\
\vdots & \vdots & \cdots\cdots & \vdots \\
X_{Q+1}^* - Z^{-1}X_Q^* & X_Q^* - Z^{-1}X_{Q-1}^* & \cdots\cdots & X_1^* - Z^{-1}X_0^*
\end{bmatrix}_{(Q+1)\times(Q+1)}
$$

$$
\times
\begin{bmatrix}
W_0 \\
W_1 \\
\vdots \\
W_V
\end{bmatrix}_{(Q+1)\times 1}
=
\begin{bmatrix}
C \\
0 \\
\vdots \\
0
\end{bmatrix}_{(Q+1)\times 1}
\tag{4.18}
$$

or equivalently,

$$
[FB][W] = [Y] \tag{4.19}
$$

4.3 Prevention of Signal Cancellation in an Adaptive Nulling algorithm

Signal cancellation is a serious problem in adaptive nulling. The problem arises when the actual direction of arrival of the signal is slightly off the assumed direction of arrival. The adaptive algorithms consider the actual signal an interferer as the direction of arrival is not exactly specified. Instead of single look-direction constraint, multiple look-direction constraints can be used to prevent signal cancellation, when the direction of arrival is not known exactly.

Correction for this effect is accomplished in least squares procedures by establishing look-direction constraints at multiple angles within the transmitter main beam extent. The multiple constraints are established by using a uniformly weighted array pattern for the same size array as that of the adaptive array under consideration. Multiple points are chosen on the nonadapted array pattern and a row is implemented in the matrix equations (4.12), (4.15), and (4.18). For each of the desired angle, and the corresponding uniform complex antenna gains are placed in the [Y] vector of (4.13), (4.16), and (4.19). Hence, for this problem the size of the matrix [U], for example, is established by the following parameters:

Q = number of look-direction constraints
L+1 = number of weights to be calculated
L-Q+1 = number of interferers that can be nulled

The first canceling equation uses data from the L+1 elements, and each successive canceling equation is shifted by one element , and therefore N - L equations are required to use the data from N+1 elements effectively. Thus there are Q-constraint equations and N – L canceling equations for the case of the forward method described by (4.12). The number of equations must equal the number of weights; therefore,

$$L = Q + N - L \tag{4.20}$$

This leads to the relationship between the number of weights, number of constraints, and number of elements:

$$N = 2L - Q \tag{4.21}$$

Similar constraints can be applied to the backward method or the forward-backward method.

4.4 Simulation Results

4.4.1 Simulation result of Forward, Backward and Forward-backward methods

A set of examples has been chosen to illustrate the direct data domain method, where use of a conventional stochastic methodology may not yield satisfactory results.

For the first example, we consider the performance of these different methods in estimating the SOI in the presence of interferers and thermal noise. We assume a desired signal (SOI) arriving from $\phi = 30^0$ impinging on a 13-element array. The array consists of elements which are omnidirectional point radiators and are considered to be spaced a half wavelength apart. We consider four interferers 20 dB stronger than the SOI signal. They are arriving from $\phi = 65^0, 85^0, -25^0$ and -50^0 respectively. In addition, we have thermal noise at each antenna element, which is assumed to have normal distribution in amplitude with zero mean, and the phase has uniform distribution between 0 and 2π. The signal to thermal noise ratio chosen is +46 dB so the standard deviation for the thermal noise is 0.01501. We consider a single snapshot of the voltages across the array at a particular instance of time. So the data that is available for analysis are:

    1) The complex voltages that consist of the signal, 4 interferers and thermal noise at each of the 13 elements

    2) The direction of arrival of the SOI, which is 30^0.

When one applies the forward method to this data based on a single snap-shot, the signal to interference plus noise ratio at the output was +52.5 dB. In this case the number of weights is seven. Similar results are obtained when the backward method is applied to the same data. It is important to note that this is a second independent estimate of the signal. A third independent estimate can be obtained by applying the forward-backward method to the same data sets. In this case, we are doubling the data and so the number of weights is nine. In this case the output signal-to-noise ratio was now +56.4 dB. The CPU time taken by the forward method on an ACER Travel Mate 240 laptop computer with a clock speed of 2.5 GHz was less than 0.045 second. The antenna beam patterns for the forward, backward and the forward-backward methods are given in figures (4.2), (4.3) and (4.4) respectively.

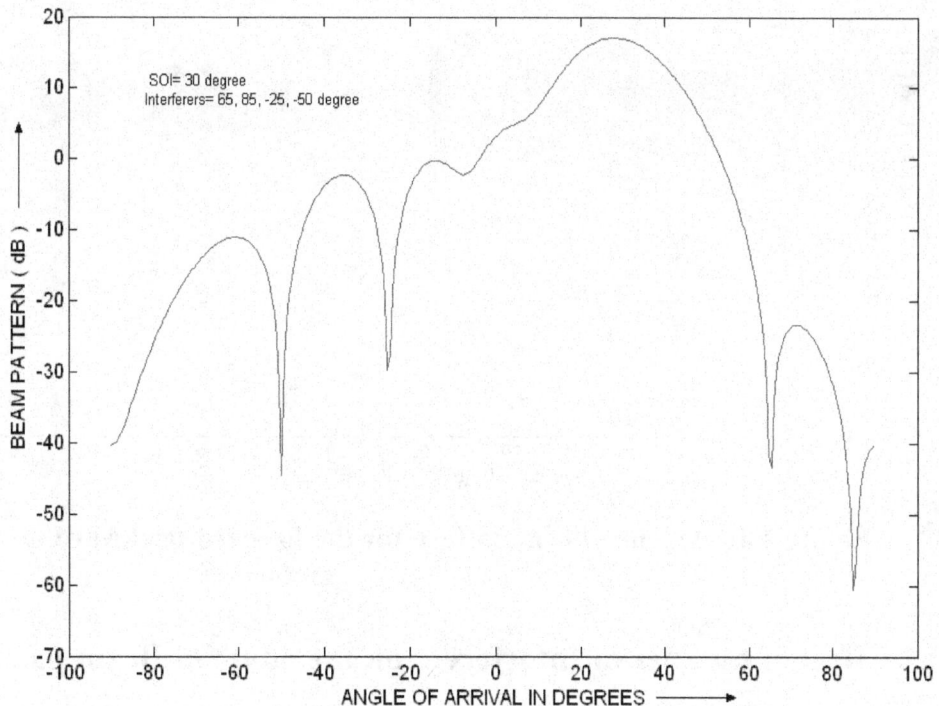

Figure 4.2:- Antenna beam pattern for the forward method.

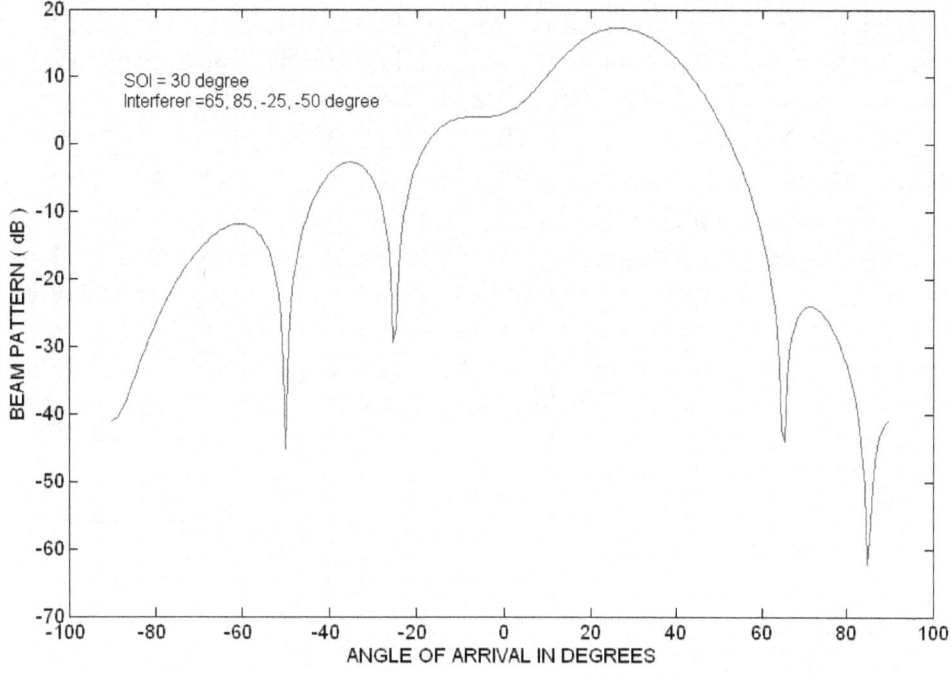

Figure 4.3:- Antenna beam pattern for the backward method.

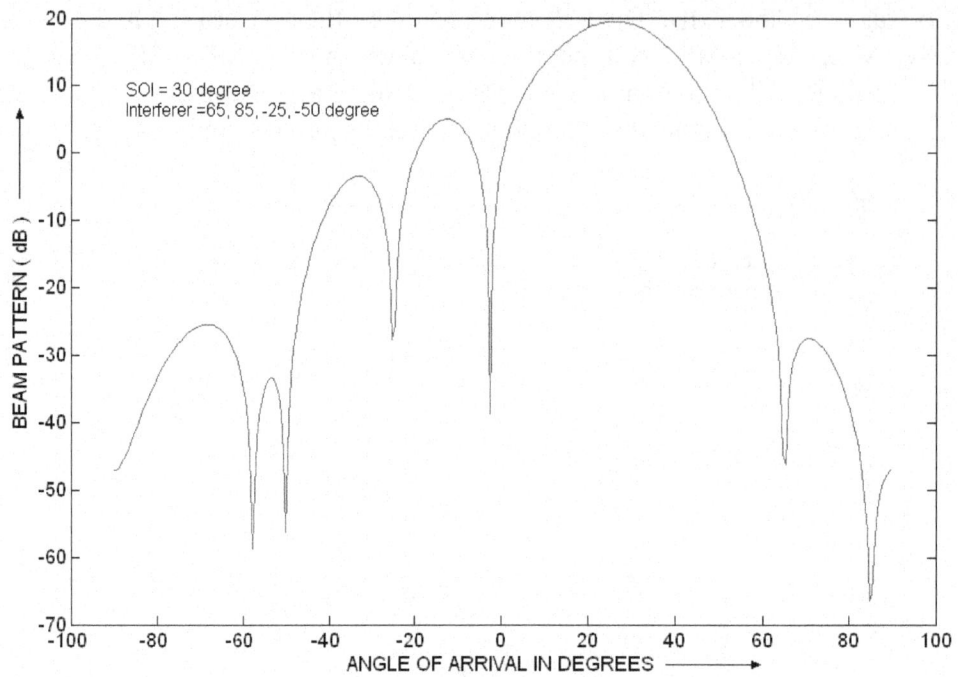

Figure 4.4:- Antenna beam pattern for the forward-backward method.

4.4.2 Effect of number of Interferers on the adapted Beam patterns

For the second example we consider the performance of the direct data domain least squares forward method when the number of interferers is more than Degrees Of Freedom of the adaptive antenna system. Consider an antenna array consisting of 13 (N+1) elements in the array. So degrees of freedom is 6 (DOF=N/2). So we can form one main beam and can null 5 interferers properly. For this example suppose desired signal and interfering signals are tone signals at 1 GHz

We consider 5 interferers 20 dB stronger than desired signal. They are arriving from $\phi = 65^0, 80^0, -15^0$, -30^0 and -45^0. In addition, we have zero mean thermal noise at each antenna element, having signal-to-thermal noise ratio +56 dB. The direction of arrival of desired signal is 30 degree and assumed to be known. Application of the forward method to this data based on a single snapshot generates the beam pattern as shown in figure (4.5). From figure (4.5) it is clear that, as number of interferers plus desired signal is equal to 6, which is equal to DOF, the nulls were placed at correct positions in the beam pattern. And Output SINR was above 66 dB.

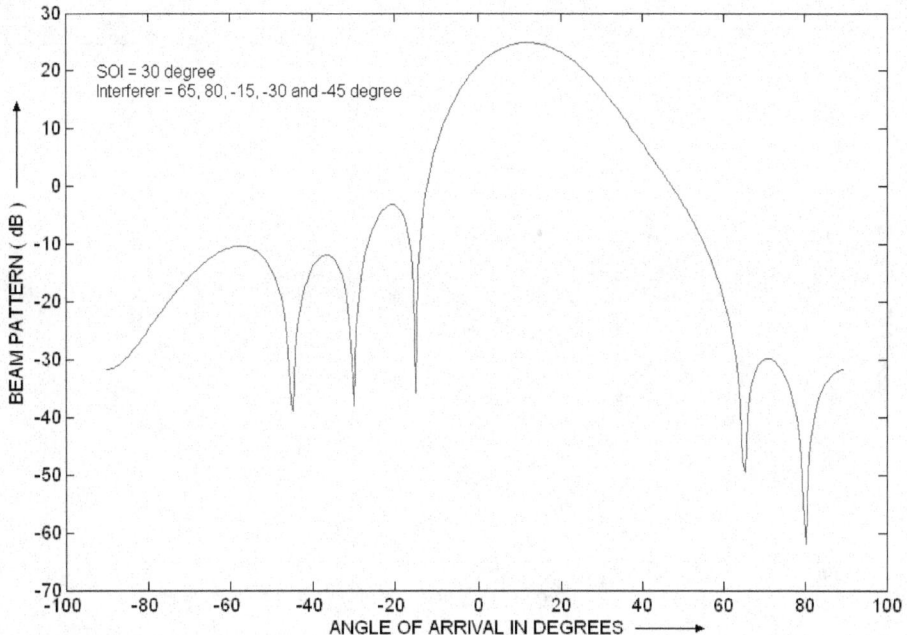

Figure 4.5:- Antenna beam pattern for the 5 interferers and SOI

Now suppose we add one more interferer, whose angle of arrival is -70 degree and is 30 dB stronger than SOI. Then application of forward method has generated beam pattern as shown in figure (4.6). Which show that null at an angle 80 degree in the beam pattern has been disappeared and null at an angle -70 degree has been formed slightly.

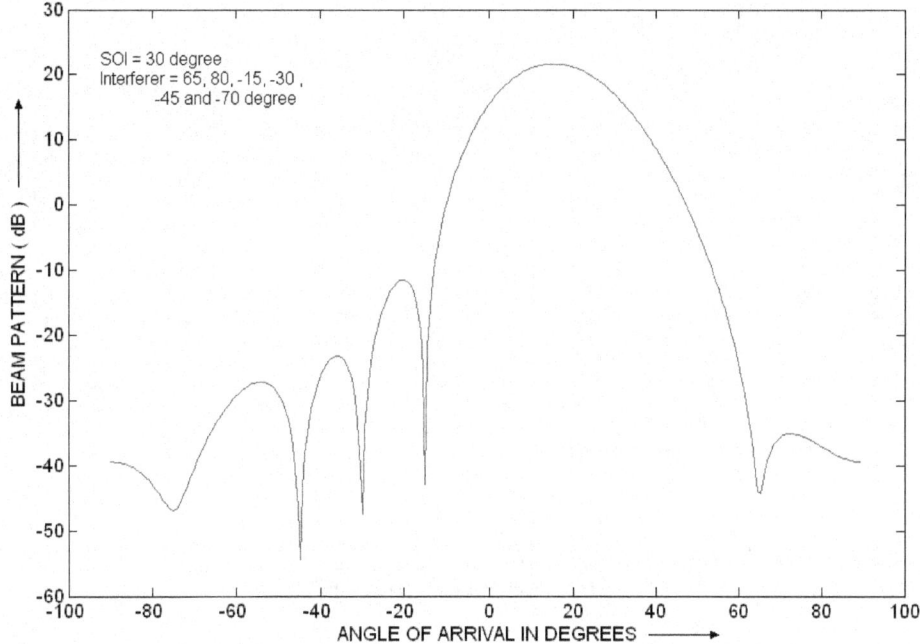

Figure 4.6:- Antenna beam pattern for the 6 interferers and SOI.

Now suppose, if strength of interferer at an angle -70 degree is increased from +30 dB to +40 dB above desired signal strength and DOAs of other interferers and SOI are unchanged. Then application of forward method has generated simulated beam pattern as shown in figure (4.7).

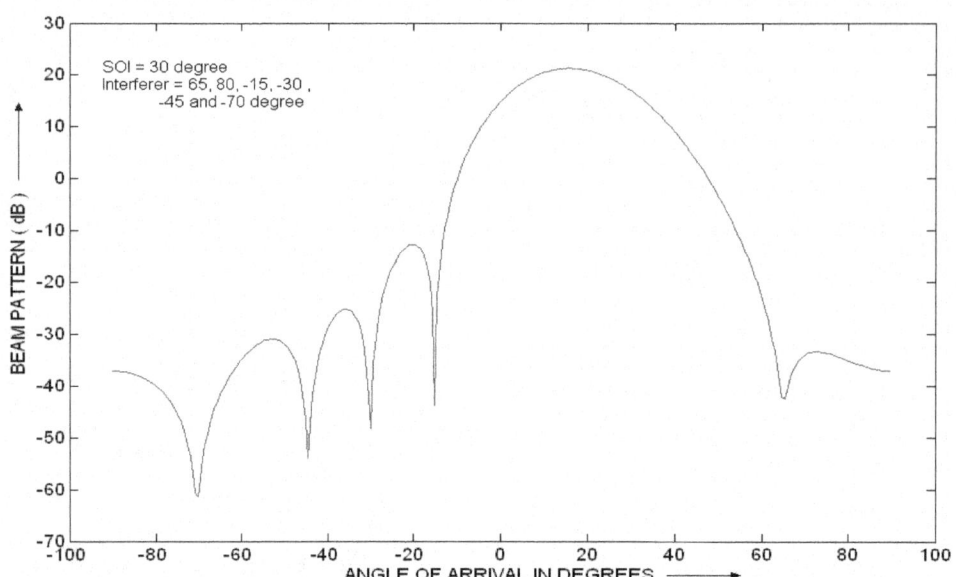

Figure 4.7:- Antenna beam pattern for the 6 interferers and SOI. (40 dB interferer at - 70 degree)

Figure (4.7) shows that null at 80 degree was disappeared and new null at -70 degree was formed. So in all only 5 nulls are formed in the beam pattern even though there are 6 interferers, as DOF (Degree Of Freedom) is 6 (one SOI + 5 interferers).

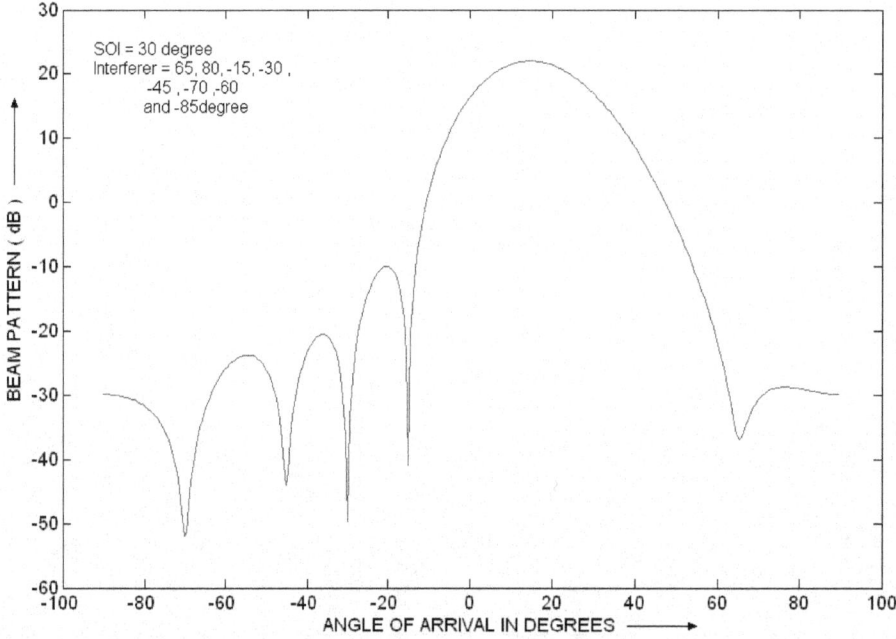

Figure 4.8:- Antenna beam pattern for the 8 interferers and one SOI

In the above scenario, if suppose two more interferes are added. The DOAs are -60 and -80 degree respectively. The strength of these two interferers is 6 dB below SOI. The simulated beam pattern for this case is shown in figure (4.8). Simulation result shows that nulls were not changed, this is because two new interferers have strength - 6 dB below SOI, whereas 5 interferers were 20 dB stronger than SOI and one interferer is 40 dB stronger than SOI. So nulls are formed at the angles, where strong interferers are present

4.4.3 Effect of Thermal Noise on the Adapted Beam Pattern

The Thermal or Gaussian noise has no effect on the position of nulls in the adapted beam pattern. However, the simulation results show that, thermal noise has effect on the sidelobe levels. If we keep positions of interferers and their magnitudes constant, and calculate adaptive weights at different time instances using corresponding voltage snapshots, then the calculated set of weights differ from one time instant to another time instant. However, the set of weights calculated have the ability to place the nulls in the direction of interferers and more gain in the direction of SOI. So if there is no interferer arrives from the sidelobe positions, then change in sidelobe level has no effect on the reconstructed desired signal.

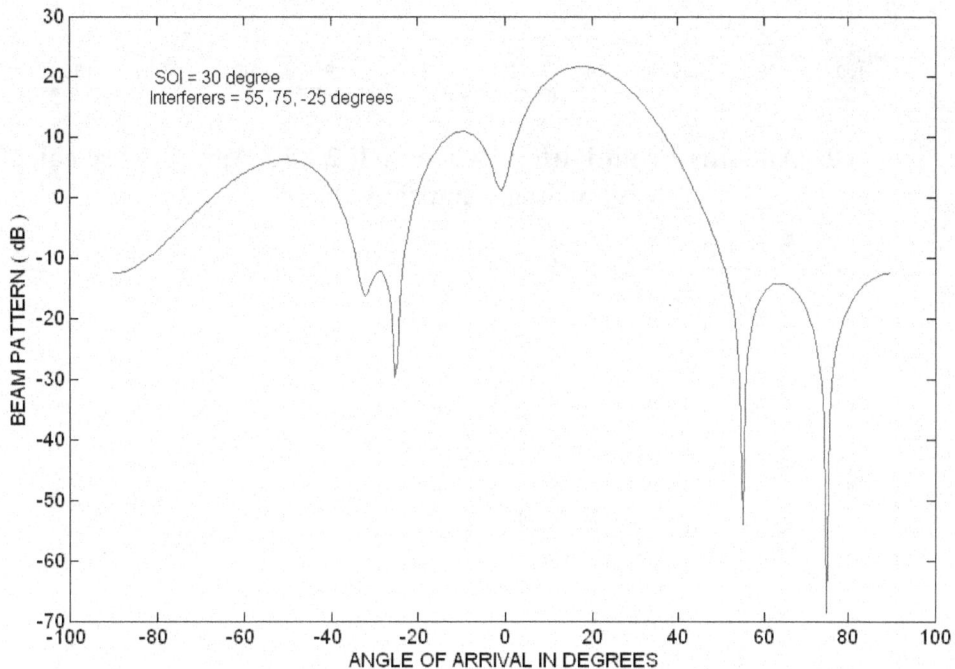

Figure 4.9:- Antenna Beam Pattern when SNR 23 dB. Weights are calculated using voltage snapshot at .1 nS

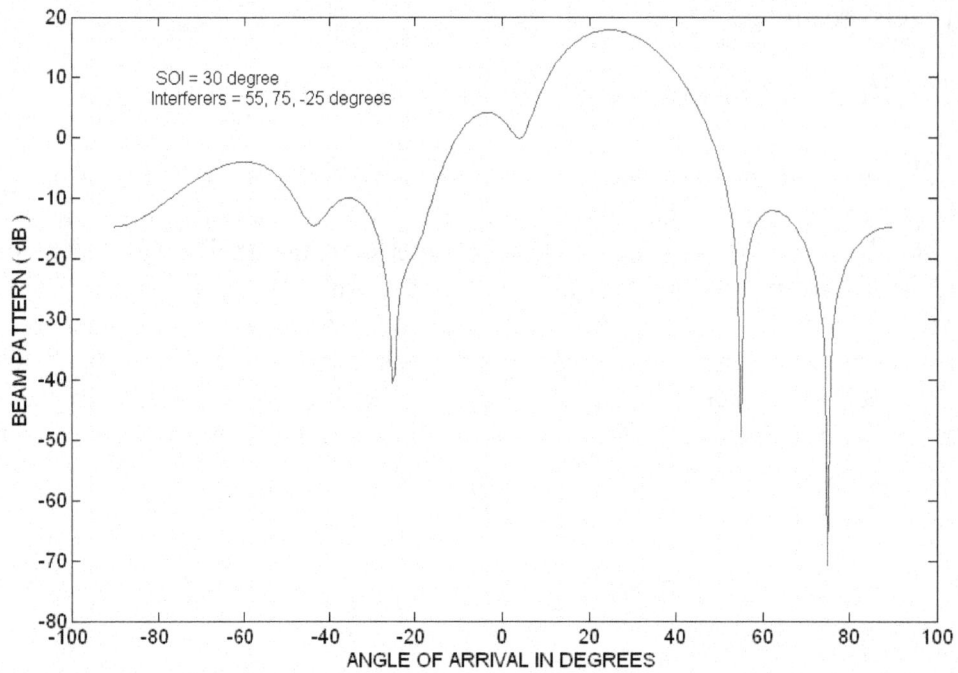

Figure 4.10:- Antenna Beam Pattern when SNR 23 dB. Weights are calculated using voltage snapshot at .2 nS

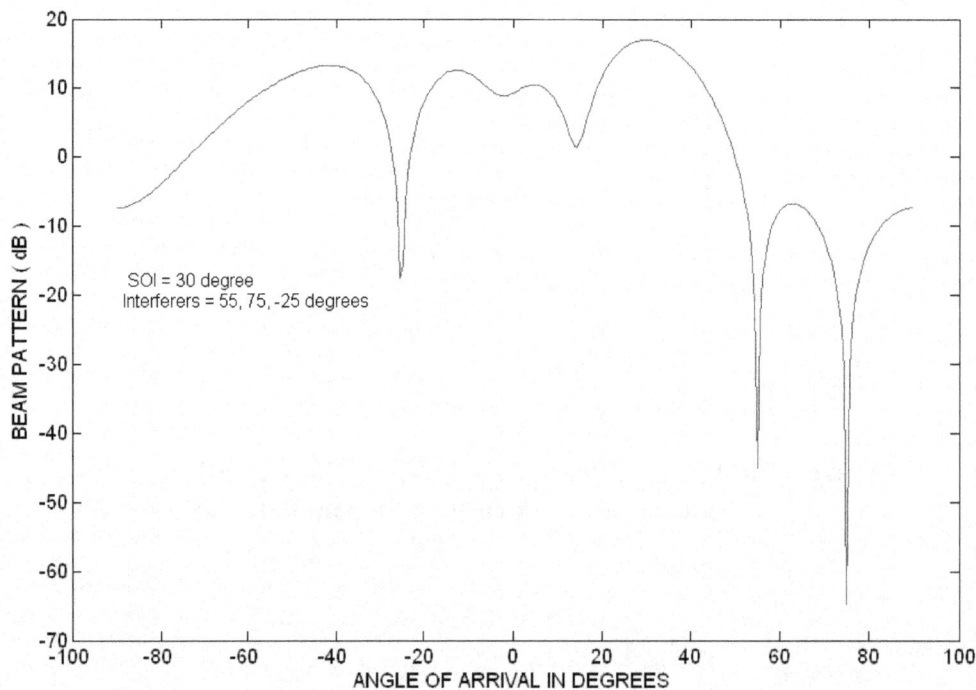

Figure 4.11- Antenna Beam Pattern when SNR 43 dB. Weights are calculated using voltage snapshot at .1 nS

The Figure (4.9), (4.10) and (4.11) illustrate the effect of thermal noise on the adapted beam pattern. In figure (4.9) and (4.10), Signal to thermal noise ratio is kept constant at +23 dB. Also three interferers +34 dB stronger than SOI, arrive from 55, 75 and -25 degree respectively. SOI arrives from 30 degree. The simulation is obtained for two different time instances 0.1 and 0.2 nanoseconds respectively, by

83

keeping strengths and DOAs of interferers and SOI constant. So from two simulation results given in Figure (4.9) and (4.10), it is clear that only sidelobe levels are changed. Figure (4.11) shows the simulation result, when Signal to thermal noise ratio is +43 dB.

4.4.4 Effect of number of interferers on the output SINR ratio

In the section (4.3.2), we have seen the effect of number of interferers on the adapted beam pattern. Now we will see the effect of number of interferers on the output SINR ratio. For this case we consider linear antenna array consisting 13 omnidirectional point sources and spaced $\lambda/2$ apart. Mutual coupling is not taken into account. Signal to thermal noise ratio is kept constant at +63 dB. The desired signal and all the interferers are tone signals at 1 GHz.

Note that for 13 element linear arrays, forward method gives Degree of Freedom equal to 6. So we will analyze the effect of number of interferers for less than 6 and greater than 6 cases. The following simulation results are obtained using MATLAB 6.5 for forward method adaptive nulling algorithm.

The output SINR (Signal to Interference plus Thermal Noise ratio) is estimated as

$$SINR = 20\log_{10}\left|\frac{\alpha_{true}}{\alpha_{est} - \alpha_{true}}\right| \tag{4.22}$$

Where,

α_{est} = estimated value of desired signal

α_{true} = true value of desired signal

Table 4.1: 2 Interferer cases (DOA of SOI is 70 degree)

	Strength of Interferer relative to SOI in (dB)	DOA in degree	Average Output SINR (dB)
Interferer 1	+20	45	**69.97**
Interferer 2	+20	20	

Table 4.2: 4 Interferer cases (DOA of SOI is 70 degree)

	Strength of Interferer relative to SOI in (dB)	DOA in degree	Average Output SINR (dB)
Interferer 1	+20	45	
Interferer 2	+20	20	**69.93**
Interferer 3	+20	0	
Interferer 4	+20	-15	

Table 4.3: 5 Interferer cases (DOA of SOI is 70 degree)

	Strength of Interferer relative to SOI in (dB)	DOA in degree	Average Output SINR (dB)
Interferer 1	+20	45	
Interferer 2	+20	20	
Interferer 3	+20	0	**69.75**
Interferer 4	+20	-15	
Interferer 5	+20	-35	

Table 4.4: 6 Interferer cases (DOA of SOI is 70 degree)

	Strength of Interferer relative to SOI in (dB)	DOA in degree	Average Output SINR (dB)
Interferer 1	+20	45	
Interferer 2	+20	20	
Interferer 3	+20	0	
Interferer 4	+20	-15	**57.19**
Interferer 5	+20	-35	
Interferer 6	+20	50	

Table 4.5: 7 Interferer case (DOA of SOI is 70 degree)

	Strength of Interferer relative to SOI in (dB)	DOA in degree	Average Output SINR (dB)
Interferer 1	+20	45	
Interferer 2	+20	20	
Interferer 3	+20	0	
Interferer 4	+20	-15	30.17
Interferer 5	+20	-35	
Interferer 6	-20	50	
Interferer 7	-20	-70	

Table 4.6: 7 Interferer case (DOA of SOI is 70 degree)

	Strength of Interferer relative to SOI in (dB)	DOA in degree	Average Output SINR (dB)
Interferer 1	+20	45	
Interferer 2	+20	20	
Interferer 3	+20	0	
Interferer 4	+20	-15	16.19
Interferer 5	+20	-35	
Interferer 6	-6	50	
Interferer 7	-6	-70	

Table 4.7: 7 Interferer cases (DOA of SOI is 70 degree)

	Strength of Interferer relative to SOI in (dB)	DOA in degree	Average Output SINR (dB)
Interferer 1	+20	45	
Interferer 2	+20	20	
Interferer 3	+20	0	
Interferer 4	+20	-15	**-8.84**
Interferer 5	+20	-35	
Interferer 6	+20	50	
Interferer 7	+20	-70	

Table 4.8: 8 Interferer cases (DOA of SOI is 70 degree)

	Strength of Interferer relative to SOI in (dB)	DOA in degree	Average Output SINR (dB)
Interferer 1	+20	45	
Interferer 2	+20	20	
Interferer 3	+20	0	
Interferer 4	+20	-15	
Interferer 5	+20	-35	**-14.78**
Interferer 6	+20	50	
Interferer 7	+20	-70	
Interferer 8	+20	-85	

From the simulation result of Table 4.2, the number of interferers present is 2. The 2 interferers are +20 dB stronger than desired signal. The average output SINR obtained was +39.02 dB. Table 4.2 shows the simulation result for 4 interferers having +20 dB stronger than SOI. The output SINR obtained was +38.53 dB.

Simulation results obtained in Table 4.1, 4.2 and 4.3 shows that, as long as number of interferers plus SOI are less than or equal to Degree Of Freedom (in this case 6), the output SINR is very good.

Table 4.4, 4.5 and 4.6 show simulation results, when number of interferers is greater than Degree Of Freedom (in this case DOF is 6). The results show that, the output SINR obtained was very poor, in Table 4.6 it is – 38.24 dB.

However, when number of interferers is greater than DOF (Degree Of Freedom), then adaptive nulling algorithm suppress the stronger interferers whose number is equal to DOF-1. The additional interferers cannot be suppressed adequately, and they degrade the output SINR depending upon their strength.

4.4.5 Effect of DOA of interferer relative to DOA of SOI on the Output SINR

For this case we consider linear antenna array consisting 13 omnidirectional point sources and spaced $\lambda/2$ apart. So the half power beam width of the array for broad side processing is approximately 7.8°. Mutual coupling is not taken into account. Signal to thermal noise ratio is kept constant at +43 dB. The desired signal and the interferer are tone signals at 1 GHz.

In this example we have SOI impinging on the array from 50 degree. To see the Effect of DOA of interferer relative to DOA of SOI on the Output SINR we consider only one interferer. The following simulation results were obtained using MATLAB 6.5 for forward method adaptive nulling algorithm based on the single snapshot of data.

Table 4.9:- Simulation result showing the Effect of DOA of interferer relative to DOA of SOI on the Output SINR

DOA of SOI (degree)	DOA of Interferer (degree)	Strength of Interferer relative to SOI	Average Output SINR
50	35	-6	67.25
50	40	-6	63.17
50	45	-6	55.33
50	47	-6	46.59
50	48	-6	38.47
50	49	-6	28.44
50	49.5	-6	14.63
50	50.5	-6	14.41
50	51	-6	27.18
50	52	-6	38.77
50	55	-6	52.72
50	60	-6	60.72
50	65	-6	65.98

The simulation results obtained in the table 4.8 shows that, when DOA of interferer is close to the DOA of desired signal (SOI), then output SINR becomes poor depending upon the intensity of the interferer. For 13 element linear array, the half-power beam width for broad side case is given by,

$$\Theta_h \approx 2\left[\frac{\pi}{2} - \cos^{-1}\left(\frac{1.391\lambda}{\pi(N+1)\Delta}\right)\right] \qquad (4.22)$$

Where,

$N+1$ = number of elements in the array.

λ = wavelength

Δ = spacing between the elements

Using equation (4.22) for 13 elements array the half power Beam width is 7.8°.

For adequate suppression of interferer, it should be away from $50^{0} \pm (7.8^{0}/2)$, when DOA of desired signal is 50^{0}. In general for adequate suppression of interfering signal, it should be spatially away from SOI, mathematically,

$$\left[\phi_S + \left(\frac{\Theta_h}{2}\right)\right] \leq \phi_I$$

or

$$\phi_I \leq \left[\phi_S - \left(\frac{\Theta_h}{2}\right)\right]$$

where,

ϕ_S = angle of arrival of SOI

ϕ_I = angle of arrival of SOI

Θ_h = half-power beam-width in degree.

4.4.6 Impact of Number of elements in the adaptive array on the Output SNR

In direct data domain algorithm, if antenna array consists of N+1 elements, then degree of freedom is (N/2) and number of weights are (N/2 +1). So in D^3LS method sub array of (N/2 +1) elements is formed for adaptive signal processing. However, in the weight calculation, output signal of all N+1 elements is used. But for estimation of desired signal sub array of (N/2 +1) elements is used.

To see the impact of number of elements in the sub array of D^3LS algorithm, consider a scenario where we have only desired signal and Gaussian or thermal noise. Interfering signals are considered to absent in the simulation. Also consider we have 13 elements linear antenna array having omnidirectional point sources spaced $\lambda/2$ distance apart. The simulation was carried out for different Input Signal to Thermal noise (SNR) ratios and average output SNR was measured. The result of simulation is given in following Table 4.10 for Forward method algorithm.

Table 4.10:- Input and Average Output SNR of Forward method adaptive nulling algorithm

Sr. Number	Input Signal to Thermal Noise Ratio (dB)	Average Output Signal to Thermal Noise Ratio (dB)
1	0	8.5
2	16.98	25.52
3	19.48	28.21
4	23.01	31.99
5	29.03	37.73
6	39.48	48.17

From the simulation result given in the Table 4.10, it is clear that, algorithm suppresses input thermal noise by a factor of approximately 8 at the output.

So in general D^3LS forward method algorithm can increase output signal to thermal noise ratio by a factor of G.

Where G = W+1

W is the number of weights in the D^3LS algorithm.

W = (N+1)/2

The number of weights W is more when an adaptive antenna system consists of more number of elements in the array. So more the number of elements in the array more is the suppression of Gaussian noise at the output of adaptive nulling system.

4.5 Conclusion

In adaptive nulling beamformer, when number of interferers plus SOI is less than or equal to degree of freedom, then nulls are formed accurately in the adapted beam pattern and output SINR is good. When number of interferers plus SOI is greater than DOF, then number of nulls equal to DOF-1 is formed accurately in the adapted beam pattern and these nulls correspond to the interferers of strong intensity, however output SINR degrades, depending upon the intensities of additional interferers. Thermal noise has no effect on the position of nulls in adapted beam pattern, but it affects side lobe level in the adapted beam pattern.

CHAPTER 5

Conjugate Gradient Method for D³LS Algorithm

5.1 Conjugate Gradient Method Algorithm

The Conjugate Gradient Method is used to solve the operator equation given by

$$[A][W] = [Y] \tag{5.1}$$

In this operator equation [A] is known square matrix, [Y] is known column vector and [W] is unknown column vector [13] [10].

The fundamental principle of the method is to select a set of vector P_i such that they are A - conjugate (or A - orthogonal) i.e.

$$\left(AP_i, AP_j \right) = 0 \qquad for \quad i \neq j \tag{5.2}$$

Equation (5.2) guarantees that the method will converge in a finite number of steps. Let the residual at the ith iteration be defined by

$$R_i = Y - AW_i \tag{5.3}$$

and let W_i denotes the approximate solution of (5.1) obtained at the ith iteration. This has been obtained by minimizing the functional.

$$F(W_i) = \left\| R_i \right\|^2 \tag{5.4}$$

Let
$$G_i = A^* R_i$$

Where, G_i is the proportional to the gradient of the functional $F(W_i)$ and A^* is complex conjugate of A.

The conjugate gradient method for this case starts with an initial guess W_0 and defines

$$R_0 = Y - AW_0 \tag{5.5}$$

$$P_0 = G_0 = A^* R_0 \tag{5.6}$$

for I = 0,1, 2,….,let

$$a_i = \frac{\|G_i\|^2}{\|AP_i\|^2} \tag{5.7}$$

$$W_{i+1} = W_i + a_i P_i \tag{5.8}$$

$$R_{i+1} = R_i - a_i AP_i \tag{5.9}$$

$$G_{i+1} = A^* R_{i+1} \tag{5.10}$$

$$P_{i+1} = G_{i+1} + b_i P_i \tag{5.11}$$

$$b_i = \frac{\|G_{i+1}\|^2}{\|G_i\|^2} \tag{5.12}$$

The equations above are applied in an iterative fashion until the desired error criterion for the residual is satisfied.

$$R_i = [Y] - [A][W_i]$$

In this case the error criterion is defined by

$$\frac{\|[R]_i\|}{\|[Y]\|} = \frac{\|[Y] - [A][W]_i\|}{\|[Y]\|} \leq 10^{-6} \tag{5.13}$$

The iterative procedure is stopped when the criterion defined above is satisfied.

The characteristic of iteration (5)-(13) is such that W_i converges to the exact solution W for all initial guesses. The error is reduced at the each iteration. This is because

$$F(W_i) - F(W_{i+1}) = \frac{|(R_i, AP_i)|^2}{(AP_i, AP_i)} \tag{5.14}$$

$$= a_i^2 (AP_i, AP_i) \text{ is always positive}$$

$\|R\|^2$ is the least possible in 'i' steps by utilizing any iterative scheme of the form.

Because $\|R_i\|^2$ is minimized this particular method essentially gives a sequence of least squares solutions to $[A][W] = [Y]$ such that at the each iteration the estimate W_i gets closer to W.

The conjugate gradient method converges to the minimum norm solution even for the singular case. Here lies the strength of the iterative method as it guarantees convergence even for the singular problem.

Once the system adapts itself to the signal, it needs the only one additional iteration per each new signal sample to obtain an acceptable solution.

5.2 Numerical Example

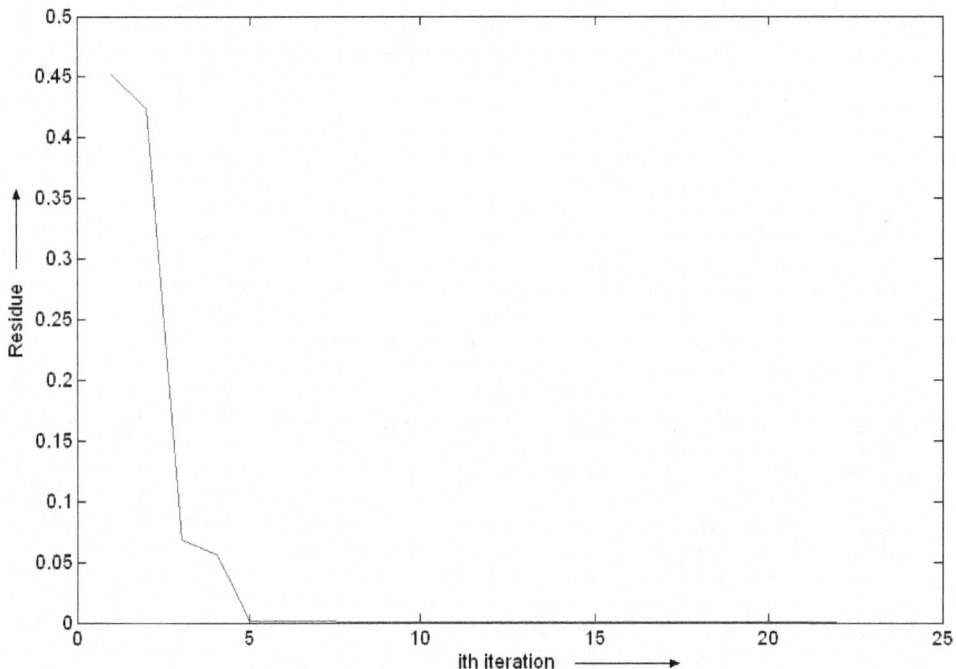

Figure 5.1:- The plot of iterations versus error (residue) generated during execution of Conjugate Gradient algorithm.

In our first numerical example we consider a single tone signal given as follows.

$$A_s = 2.00 \quad f_s = 1 \text{MHZ} \quad \phi_s = 0^0$$

And the two interferers are given as follows.

$$I_1 = 1.00 \text{ volt} \qquad f_{i1} = 1 \text{ MHz} \qquad \phi_{i1} = 61^0$$
$$I_2 = 1.25 \text{ volt} \qquad f_{i2} = 1 \text{ MHz} \qquad \phi_{i2} = 88^0$$

Where A, I denotes amplitude, f denotes frequency and ϕ denotes angle from broadside of the arrays. The sampling frequency at which the voltage measured at each antenna element is $10 f_s$. The separation between array elements is equal to $\frac{\lambda}{2}$.

The results obtained are as follows. The error $F(W_i)$ for the each iteration was obtained as:

Table 5.1: Error generated during ith iteration in conjugate gradient method

i^{th} iteration	$F(W_i)$
1	0.4518
2	0.4231
3	0.0679
4	0.0564
5	0.0011
⋮	⋮
20	1.1713 E-5
21	1.7409 E-6
22	8.2014 E-7

The residue or error $F(W_i)$ generated during the each iteration is plotted in Figure (5.1.) For this example these iterations are required only at the beginning before adaptation starts. Once the system adapts itself to the signal it needs only one or two additional iterations per each new signal sample to obtain an acceptable solution.

The first ten samplings of the signal and the system outputs are compared as follows

Signal	Output
1.0000+0.0000i	1.0000+0.0000i
0.8090+0.5878i	0.8091+0.5878i
0.3090+0.9511i	0.3090+0.9511i
-0.3090+0.9511i	-0.3090+0.9511i
-0.8090+0.5878i	-0.8090+0.5878i
-1.0000+0.0000i	-1.0000-0.0000i
-0.8090-0.5878i	-0.8090-0.5878i
-0.3090-0.9511i	-0.3090-0.9511i
0.3090-0.9511i	0.3090-0.9511i
0.8090-0.5878i	0.8090-0.5878i

The weight vector $[W]$ for this single tone signal example is

W_1 =0.8311-0.1468i

W_2 =1.1829+0.1126i

W_3 =0.9183-0.0610i

W_4 =1.1357-0.0000i

W_5 =0.9183+0.0610i

W_6 =1.1829-0.1126i

W_7 =0.8311+ 0.14

The total CPU time taken for the above result is 0.01 second (10 mS) on Acer Travel mate 40 model with clock speed of 2.5 GHz, when number iterations is 22. Once the system adapts itself to the signal it needs only one or two iterations per each new signal sample to obtain an acceptable solution and CPU time can reduce to (0.9 mS).

CHAPTER 6

Estimation of the Direction of Arrival Using the Matrix Pencil Method

6.1 INTRODUCTION

High resolution DOA estimation algorithms have many important applications in wireless communications and spatial beamforming. The class of noise subspace algorithms, such as, ESPRIT, MUSIC, is some of the most popular algorithms proposed for DOA estimation. These algorithms separate the noise and signal subspaces based on an eigenvalue decomposition of spatial covariance matrix. This matrix is estimated by averaging over several snapshots. In the case of Root-MUSIC, the eigenvectors of the noise subspace are used to form a complex polynomial whose roots correspond to the signal DOA.

Given enough snapshots, MUSIC type algorithms yield fairly accurate results. However, two shortcomings limit their use in real time applications. First, the computation complexity of noise subspace algorithms is usually very high since covariance matrix and root estimation are very expensive operations. Sophisticated hardware is required to carry these operations which increases the cost of manufacture and decreases the ability to run in real time. Second, several snapshots are required for accurate estimate of the noise subspace algorithms.

In radar and other signal processing applications, the Matrix Pencil (MP) algorithm has been shown to provide accurate DOA estimates with a single snapshot [2] and extremely low computation burden. However, a major limitation in applying this technique to the CDMA case is that, given an array of (N+1) elements, MP can generate an accurate estimate only when

$$P \leq \frac{N+1}{2}$$

where P is the number of source presented.

The advantage of using MP is that the computation complexity is much lower than that of noise subspace algorithms because MP does not have to estimate any spatial covariance matrix or to find any roots of a polynomial. Besides, MP works well even with a single snapshot which makes it very attractive to real time applications.

6.2 Direction of Arrival Estimation in One Dimension using Matrix Pencil Method

The objective is to estimate the directions of arrival (DOA) of several signals using the Matrix Pencil (MP) in one dimension. If we have a uniformly spaced array of omnidirectional isotropic point sensors located along the x-axis and distance between any two of them is Δx. Mutual coupling between antenna elements is not taken into account. One can write the voltage X(n) induced in each of the n antenna elements, for n = 0, 1, 2,………,N-1, as

$$X(n) = \sum_{p=0}^{P} A_p \ \exp(\frac{j2\pi \ \Delta x \ n \ \sin\phi_p}{\lambda}) \qquad n = 0, 1, 2, \dots\dots\dots, \text{N-1} \qquad (6.1)$$

$$X(n) = \sum_{p=0}^{P} A_p \ z_p^{\ n} \qquad\qquad\qquad\qquad (6.2)$$

Where,

$$z_p = \exp(\frac{j2\pi \ \Delta x \ \sin\phi_p}{\lambda}) \qquad\qquad\qquad\qquad (6.3)$$

A_p = complex amplitude of the pth signal induced at the antenna element

ϕ_p = direction of arrival of pth signal

X(n) represents the voltages measured at each of the N elements of the linear array and is assumed to be known. For DOA estimation, the goal is to estimate z_p and P

A_p is ignored. Consider the following matrix,

$$[X] = \begin{bmatrix} X_0 & X_1 & \cdots\cdots & X_L \\ X_1 & X_2 & \cdots\cdots & X_{L+1} \\ \vdots & \vdots & \cdots\cdots & \vdots \\ X_{N-L} & X_{N-L+1} & \cdots\cdots & X_{N-1} \end{bmatrix}_{(N-L)\times(L+1)} \qquad (6.4)$$

where L is called pencil parameter,
Define the following two matrices,

$$[Y_1] = \begin{bmatrix} X_0 & X_1 & \cdots\cdots & X_{L-1} \\ X_1 & X_2 & \cdots\cdots & X_L \\ \vdots & \vdots & \cdots\cdots & \vdots \\ X_{N-L-1} & X_{N-L} & \cdots\cdots & X_{N-2} \end{bmatrix}_{(N-L)\times(L)} \qquad (6.5)$$

$$[Y_2] = \begin{bmatrix} X_1 & X_2 & \cdots\cdots & X_L \\ X_2 & X_3 & \cdots\cdots & X_{L+1} \\ \vdots & \vdots & \cdots\cdots & \vdots \\ X_{N-L} & X_{N-L+1} & \cdots\cdots & X_{N-1} \end{bmatrix}_{(N-L)\times(L)} \qquad (6.6)$$

These two matrices can be written as

$$Y_1 = Z_1 R Z_0 Z_2 \qquad\qquad\qquad\qquad (6.7)$$
$$Y_2 = Z_1 R Z_2 \qquad\qquad\qquad\qquad (6.8)$$

Where,

$$[Z_1] = \begin{bmatrix} 1 & 1 & \cdots\cdots & 1 \\ z_1 & z_2 & \cdots\cdots & z_P \\ \vdots & \vdots & \cdots\cdots & \vdots \\ z_1^{N-L-1} & z_2^{N-L-1} & \cdots\cdots & z_P^{N-L-1} \end{bmatrix}_{(N-L)\times(P)} \tag{6.9}$$

$$[Z_2] = \begin{bmatrix} 1 & z_1 & \cdots\cdots & z_1^{L-1} \\ 1 & z_2 & \cdots\cdots & z_2^{L-1} \\ \vdots & \vdots & \cdots\cdots & \vdots \\ 1 & z_P & \cdots\cdots & z_P^{L-1} \end{bmatrix}_{(P)\times(L)} \tag{6.10}$$

$$Z_0 = diag[z_1 \quad z_2 \quad \cdots\cdots \quad z_P] \tag{6.11}$$

$$R = diag[R_1 \quad R_2 \quad \cdots\cdots \quad R_P] \tag{6.12}$$

Note that if the entries of the matrix Z_0 can be estimated, an estimation of the DOA may be obtained using equation (6.3).
Consider the following matrix pencil,

$$Y_1 - \lambda Y_2 = Z_1 R [Z_0 - \lambda I] Z_2 \tag{6.13}$$

Therefore $\lambda = z_p$, p = 1, 2,.........,P would be the eigenvalue of the generalized eigenvalue problem $Y_1 - \lambda Y_2$

$$\phi_p = \sin^{-1}\left(\frac{c \times \cos^{-1}(real(z_p))}{\omega_p \times \Delta x}\right) \tag{6.14}$$

where
    c = speed of light 3×10^8 m/s
    ω_p = frequency of pth source in rad/sec

6.3 Validity of eigenvalues in Matrix Pencil

Validity of eigenvalues can be easily understood with the simulation example. Consider an array consists of 13 elements, so the Degree Of Freedom is 6 for Direct Data Domain Least squares algorithms. The Matrix Pencil algorithm based on single snapshot of data, can estimate DOA of 6 signals, when number of antenna elements is 13. Now suppose number of spatially distributed signals impinging on the array is 3 (less than 6). Then Matrix Pencil algorithm always provides 6 eigenvalues for 13 elements linear array, for arbitrary number of impinging signals on the array.

However, as number of impinging signals on the array is three, then out of 6 eigenvalues only three eigenvalues would be valid, and DOA as well as number of impinging signals can be estimated from valid eigenvalues only.

Suppose we have 3 signals impinging on the array from $\phi_p = 10$, 35 and 50 degree respectively having Signal to Thermal noise ratio of +43 dB. The decision regarding valid eigenvalues can be taken by using the following procedure.

1) Solve $Y_1 - \lambda Y_2$, which will give a set of six eigenvalues as

Index. Number	Eigenvalue
1	-2.2108 - 0.1155i
2	1.2772 – 1.0195i
3	-0.4336 - 0.9009i
4	-0.9126 + 0.4080i
5	0.8549 + 0.5195i
6	0.1218 + 0.5948i

2) Calculate $Y_2 - \lambda Y_1$ for a new set of six eigenvalues

Index. Number	Eigenvalue
1	-0.4338 + 0.9012i
2	-0.9132 - 0.4083i
3	-0.4511 + 0.0236i
4	0.4782 + 0.3817i
5	0.8543 - 0.5191i
6	0.3304 - 1.6137i

3) Arrange complex and absolute values of eigenvalues obtained in step (1), in ascending order

Complex Eigenvalue	Absolute Eigenvalue
0.1218 + 0.5948i	0.6071
-0.9126 + 0.4080i	0.9996
-0.4336 - 0.9009i	0.9998
0.8549 + 0.5195i	1.0004
1.2772 - 1.0195i	1.6342
-2.2108 - 0.1155i	2.2138

3) Arrange absolute values of eigenvalues obtained in step (2), in descending order

Complex Eigenvalue	Absolute Eigenvalue
0.3304 – 1.6137i	1.6472
-0.9132 - 0.4083i	1.0004
-0.4338 + 0.9012i	1.0002
0.8543 - 0.5191i	0.9996
0.4782 + 0.3817i	0.6119

-0.4511 + 0.0236i	0.4517

4) Add absolute values obtained in steps (3) and (4)

Index. Number	Sum of Absolute values of eigenvalue obtained in step(3) & (4)
1	2.2543
2	2.0000
3	2.0000
4	2.0000
5	2.2461
6	2.6655

5) As shown in table of step (5) the value of summation of absolute values of eigenvalues is 2 at the Index Number 2, 3 and 4. Number of 2's obtained after adding Absolute values of eigenvalue in step (5) gives the number of signal sources P. In our example as there are three 2's in the step (5), so the number of signals impinging on the array are 3.

6) The DOA cannot be estimated from absolute value of eigenvalue. But step (5) provides index numbers of valid Eigenvalues. So use index number 2, 3 and 4 of step (5) and corresponding valid complex eigenvalues from step (3). The valid complex eigenvalues from step (3) using Index numbers obtained in step (5) are

-0.9132 - 0.4083i
-0.4511 + 0.0236i
0.4782 + 0.3817i

8) Using valid eigenvalues, obtain the DOA of signals using equation (8.14)

The three DOA values obtained in MALAB simulation using above procedure were 10.0002, 34.9821 and 50.1623 degrees respectively. The true values were 10, 35 and 50 degrees respectively. If number of signals impinging on the array is 6 then we would get six 2's in step (5), so all six eigenvalues would be valid.

In general if P spatially separated signals impinge on the array and if P is less than or equal to Degree Of Freedom of Matrix Pencil algorithm, then in step (5) we obtain number P and P Index numbers corresponding to the valid eigenvalues in step (3).

Use Index numbers obtained in step (5) to get corresponding valid complex eigenvalues in step (3).

If the number of signals P impinging on the array is greater than Degree Of freedom of Matrix Pencil algorithm, then Matrix Pencil DOA estimation algorithm based on the single snapshot of data can estimate DOA of number of signals equal to or less than the Degree Of Freedom and having stronger signal intensities.

6.4 Simulation Results

6.4.1 Simulation Results Using Ideal Omnidirectional point sources:-

Computer simulations are carried out to illustrate the performance of the direct data domain approach. We consider a 13-element linearly equispaced array whose spacing is a half wavelength. All the elements are assumed to be ideal and consist of isotropic point sources receiving in free space.

All the signals, including the desired signal and interferers, are assumed to be tone signals at same or different frequencies. All the signals impinging on the array are assumed to arrive from 90^0 elevation angle. The received signal impinging at the array is down converted and sampled at a rate 10 times the down converted frequency.

For the first example, it is assumed that the SOI arrives at the array from 85^0 to the azimuth direction. The signal is corrupted only by Gaussian noise. The signal to noise ratio (SNR) is set at different levels. A total of 1000 simulations have been used for this example.

The result of 1000 estimations is shown in figure (6.1) when SNR is 66dB.

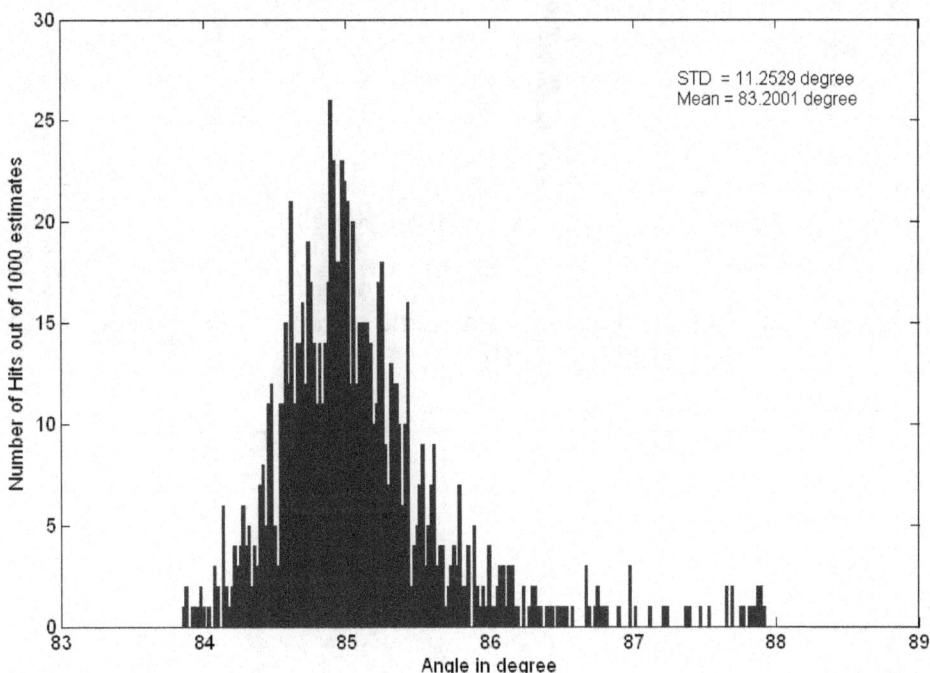

Figure 6.1:- DOA estimation of a single signal in the presence of Gaussian noise (SNR=66 dB)

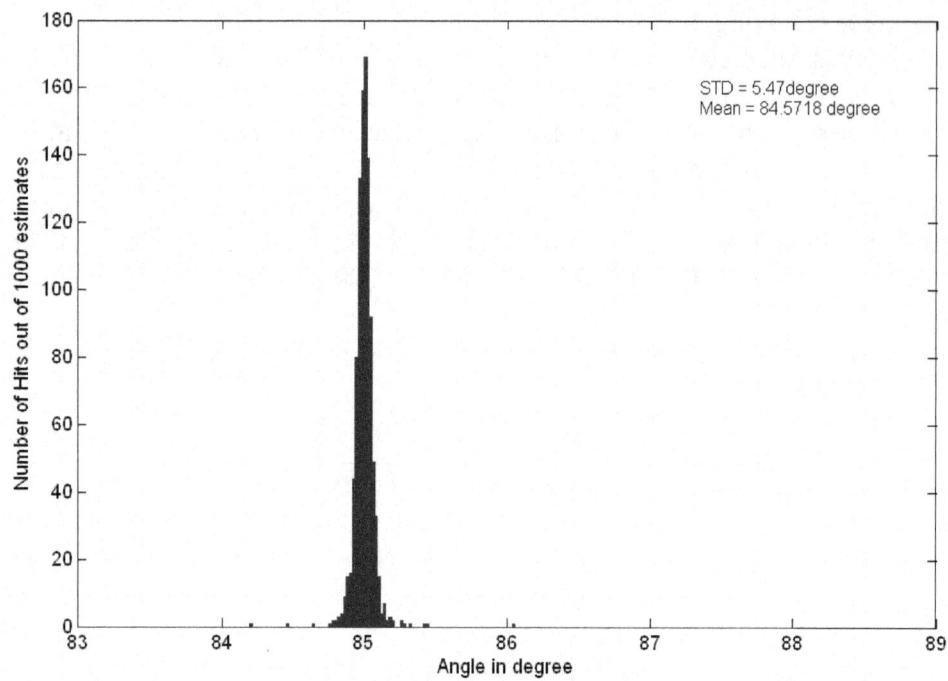

Figure 6.2:- DOA estimation of a single signal in the presence of Gaussian noise (SNR=86 dB)

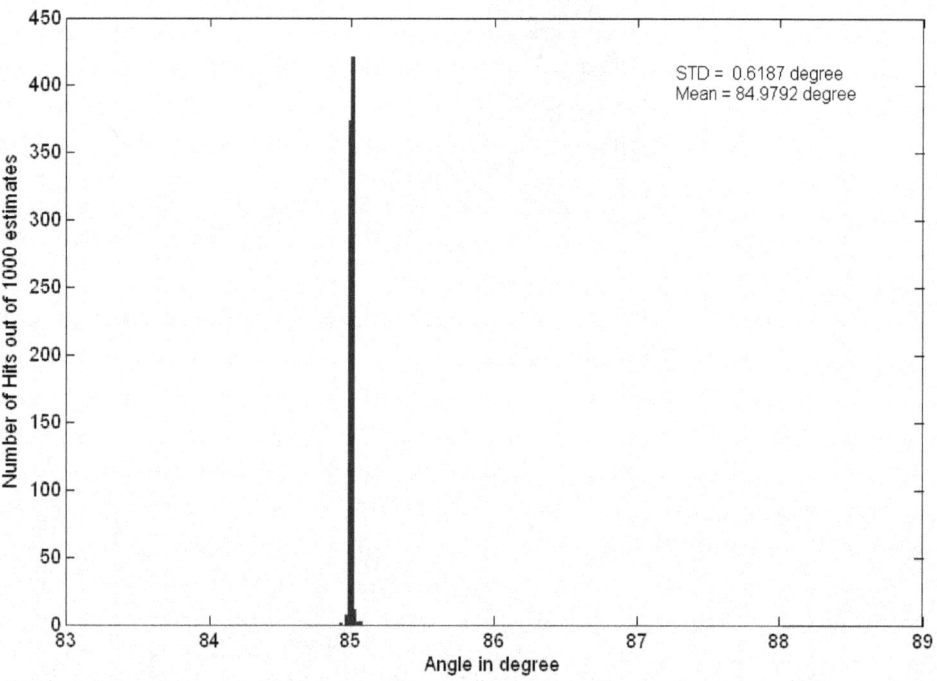

Figure 6.3:- DOA estimation of a single signal in the presence of Gaussian noise (SNR=106 dB)

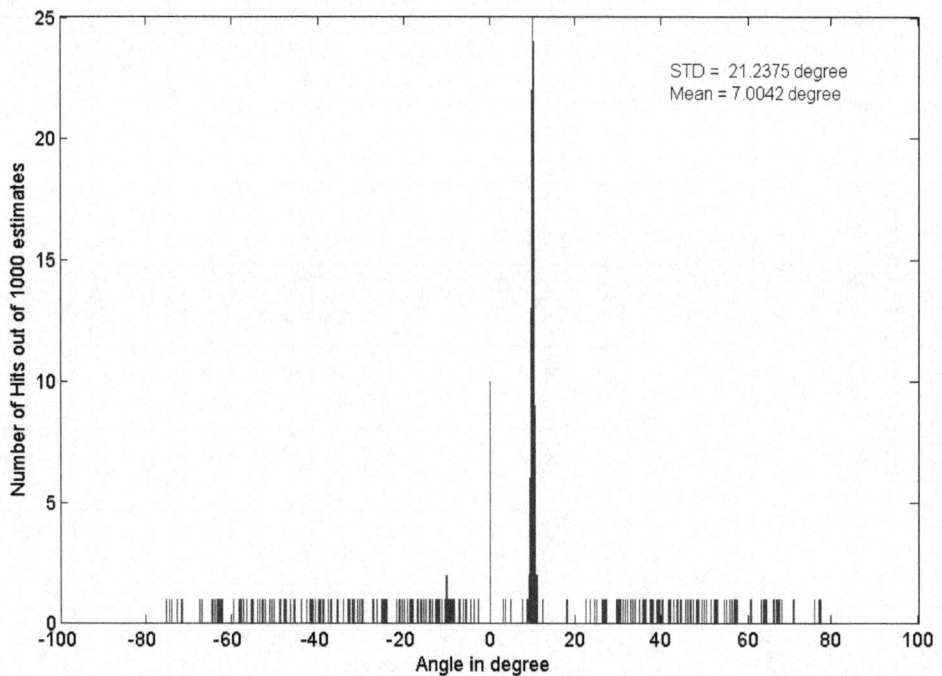

Figure 6.4:- DOA estimation of a single signal in the presence of Gaussian noise (SNR=18 dB)

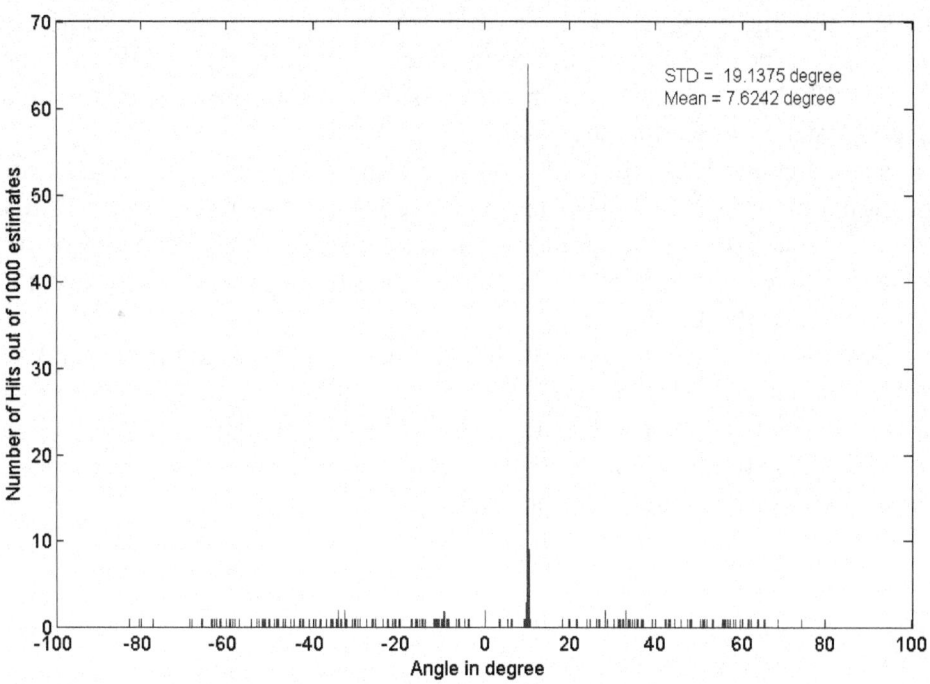

Figure 6.5:- DOA estimation of a single signal in the presence of Gaussian noise (SNR=28 dB)

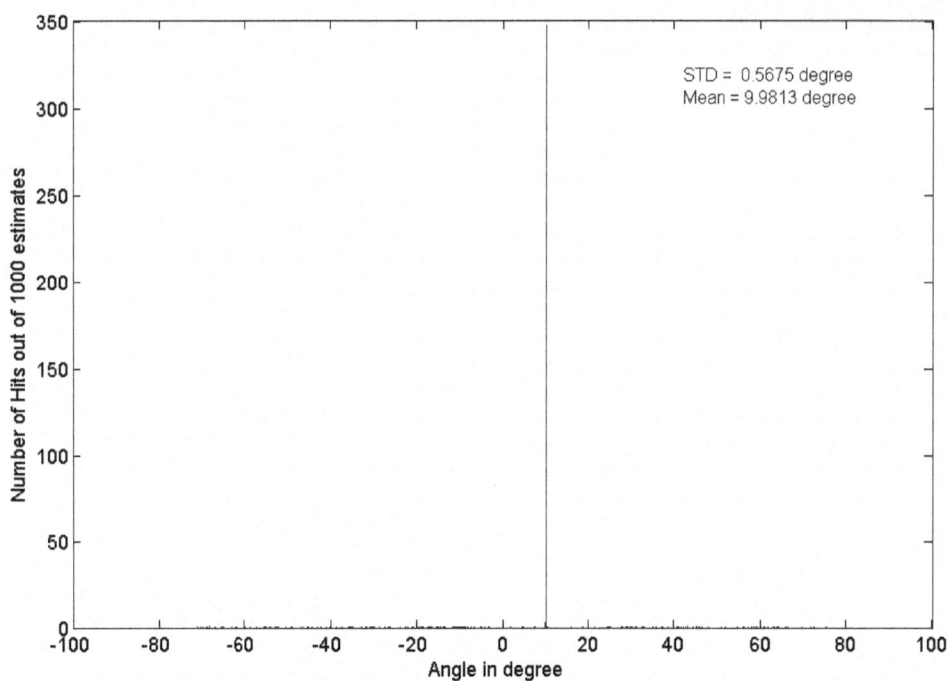

STD = 0.5675 degree
Mean = 9.9813 degree

Figure 6.6:- DOA estimation of a single signal in the presence of Gaussian noise (SNR=46 dB)

In the first example the figures (6.1), (6.2) and (6.3) shows DOA estimation of a single signal in the presence of Gaussian noise for three different SNRs 66 dB, 86 dB and 106 dB respectively. The direction of arrival of signal is held constant at 85 degree. As SNR increases, the estimation of signal approaches towards more accurate value and standard deviation decreases.

In the second example the figures (6.4), (6.5) and (6.6) shows DOA estimation of a single signal in the presence of Gaussian noise for three different SNRs 18 dB, 28 dB and 46 dB respectively. The direction of arrival of signal is held constant at 10 degree. As SNR increases, the estimation of signal approaches towards more accurate value and standard deviation decreases.

From these two examples it is clear that when signal impinge on the array from the broadside, more accurate DOA estimation is possible even at lower SNR value as compared to direction of arrival of signal from end fire. The figures (6.7) and (6.8) show the DOA estimation for three signals in the presence of Gaussian noise for two different SNRs 17 dB and 24 dB respectively.

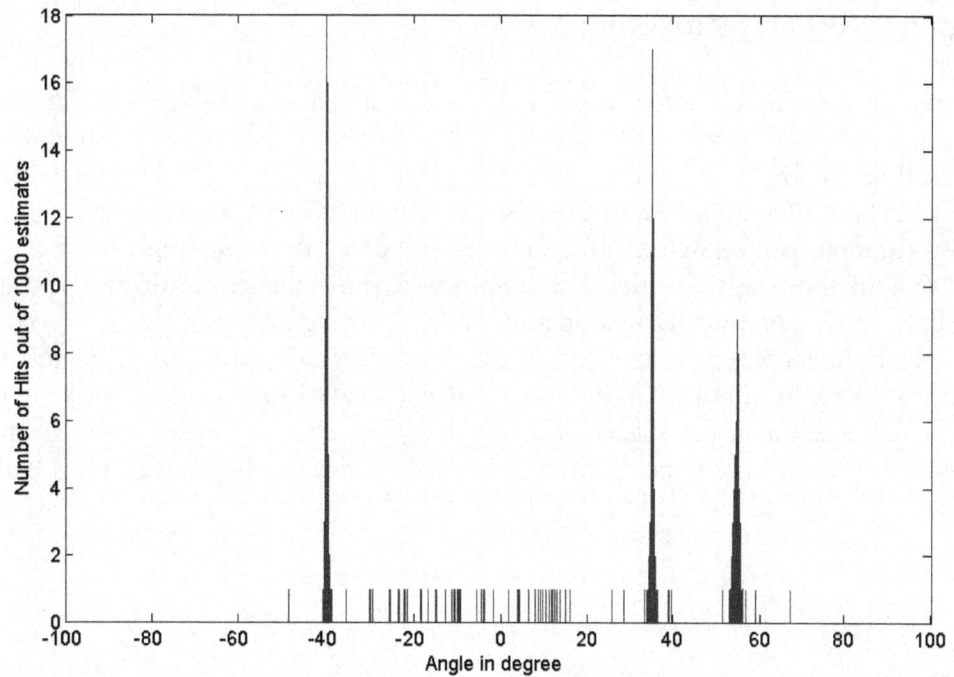

Figure 6.7:- DOA estimation of the three signals in the presence of Gaussian noise (SNR=17 dB)

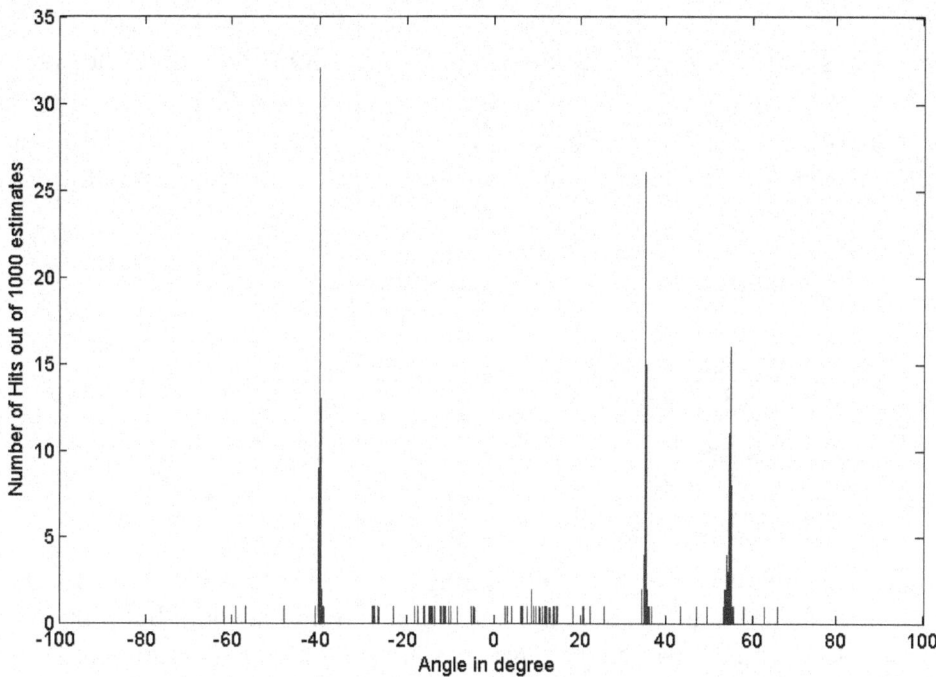

Figure 6.8:- DOA estimation of the three signals in the presence of Gaussian noise (SNR=24 dB)

102

6.4.2 Simulation Results in the Presence of Mutual Coupling   using Dipole Wire array antenna

We now present four examples illustrating the performance of the Matrix Pencil DOA estimation. First example will show the effect of presence of mutual coupling on the estimation of DOA of single signal. Second example will show the effect of compensation of mutual coupling on the DOA estimation of single signal.

Third example will show the effect of presence of mutual coupling on the estimation of DOA of three signals. Fourth example will show the effect of compensation of mutual coupling on the DOA estimation of the three signals.

For all the examples we consider an antenna array consisting of 13-elements. The elements are wire dipoles having length of half wavelength at the center frequency. The antenna elements are spaced half wavelength apart. The antenna array is analyzed using the method of moments (MM). The intensities of the signal along with their directions of arrival are used to calculate the MM voltage vector.

The equation

$$[V_{meas}] = [Z_L][I_{port}] \qquad\qquad (6.15)$$

Where $[V_{meas}]$ is a vector of measured voltages at the feeds of the array elements.

$$[I_{port}] = [Z]^{-1}[V] \qquad\qquad (6.16)$$

where $[Z]$ is MOM impedance matrix and $[V]$ is MOM voltage vector and used to find the voltages that are measured across the load at the individual feed ports. These measured voltages then serve as input to the Matrix Pencil DOA estimation algorithm. In the first scenario, no attempt is made to compensate for mutual coupling.

TABLE 6.1
Parameters defining the elements of the array.

Number of elements in array	13
Length of z-directed wires	$\lambda/2$
Radius of wires	$\lambda/200$
Spacing between wires	$\lambda/2$
Loading at the center	50 Ω

In order to see the effect of mutual coupling on the performance of DOA estimation using Matrix Pencil, the Gaussian noise is not taken into account but mutual coupling is taken into account. The single signal is impinging on the array from 0 degree direction. No attempt is made to compensate for the mutual coupling. A total of 1000 simulations have been used for this example, and result of simulation is shown in figure (6.9). The mean value of DOA estimation is 1.7269 degree, where as true value of DOA is 0 degree. The number of hits near true value is 575. So from this example it is clear that, because of mutual coupling, the estimated value is shifted from the true value.

In the second example same scenario as in first example is considered except here mutual coupling is compensated using open circuit voltage technique. A total of 1000 estimations have been used for this example, and result of simulation is shown in figure (6.10). The mean value of DOA estimation is 0.26 degree, where as true value of DOA is 0 degree. The result shows that number of hits near true value is increased from 575 to 600. Note that, compensation of mutual coupling using open circuit voltage technique reduces the mutual coupling effect but not eliminates the mutual coupling.

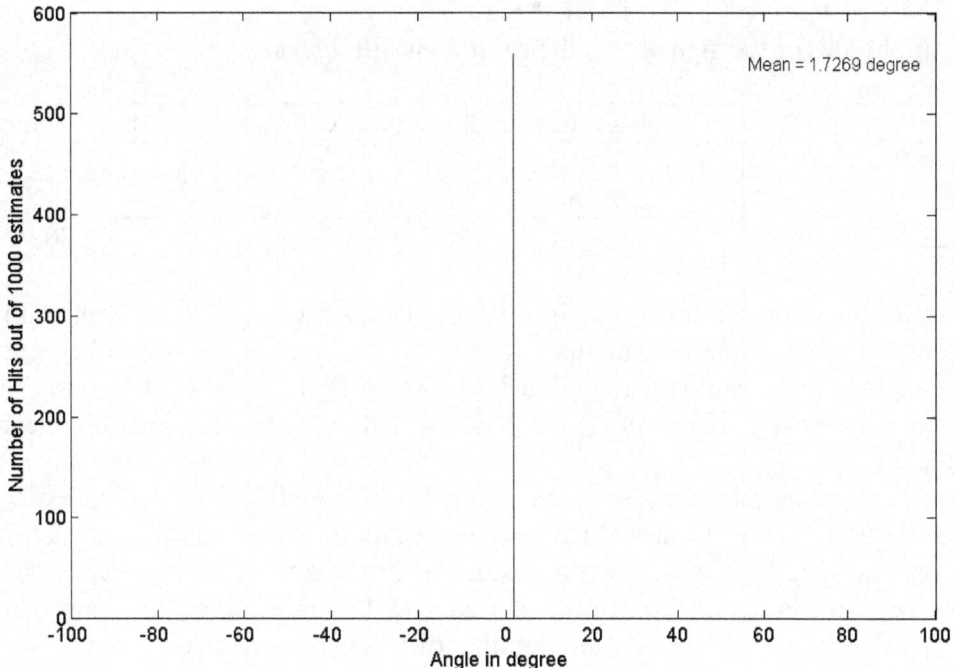

Figure 6.9:- DOA estimation of a single signal in the presence of Mutual coupling. (Gaussian noise is absent).

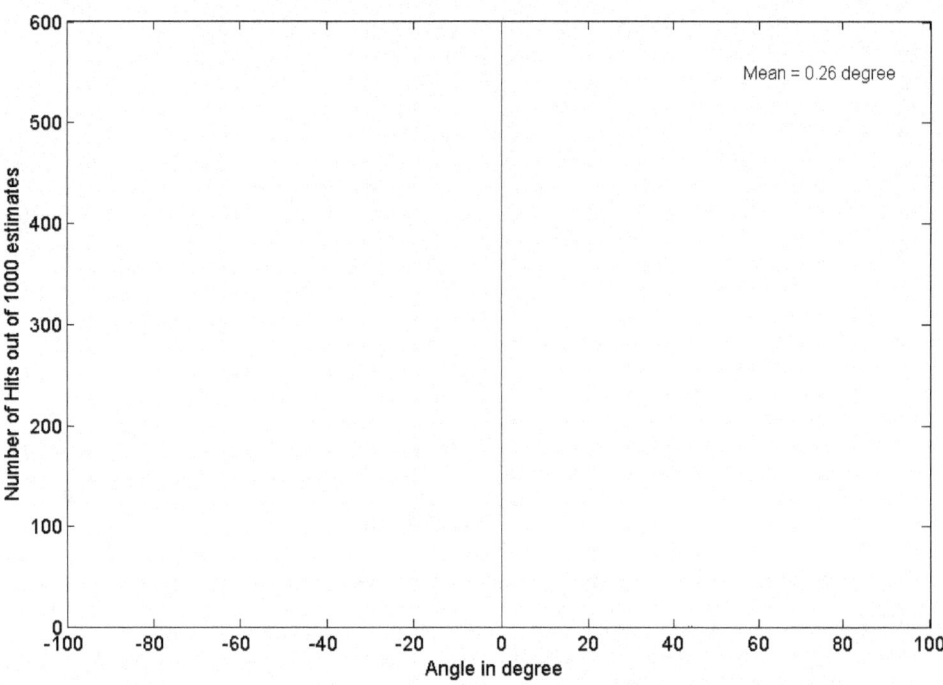

Figure 6.10:- DOA estimation of a single signal when Mutual coupling is compensated using open circuit voltage (Gaussian noise absent)

In the third and fourth examples three signals are impinging on the array , the direction of arrival and their amplitudes are listed in Table (6.2)and signal to Gaussian noise ratio is set at 47 dB.

TABLE 6.2
Amplitudes of the signals and their directions of arrival.

	Magnitude	Phase(degree)	DOA (degree)
Signal 1	2.0 V/m	0	20
Signal 2	2.0 V/m	55	45
Signal 3	2.0 V/m	35	-80

In the third example mutual coupling is taken into account. No attempt is made to compensate for the mutual coupling. A total of 1000 estimations have been used for this example, and result of simulation is shown in figure (6.11). The number of hits near the true values of DOAs is less. Moreover there are some false hits near the 0 degree.

In the fourth example same scenario as in third example is considered except here mutual coupling is compensated using open circuit voltage technique. A total of 1000 estimations have been used for this example, and result of simulation is shown in figure (6.12). The simulation result in figure (6.12) shows that, the number of hits near the true values is increased and number of false hits is reduced.

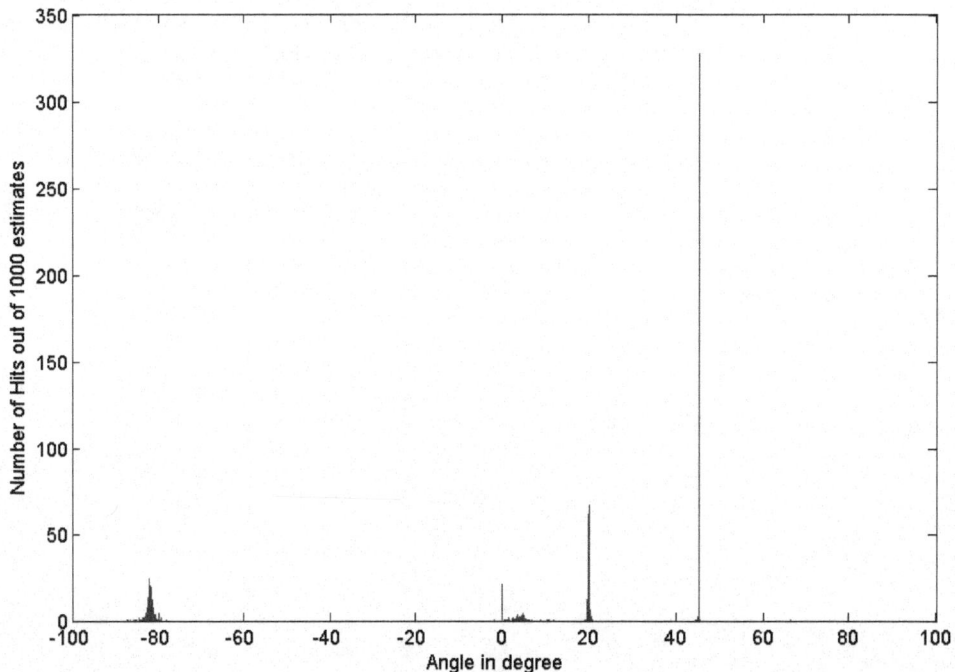

Figure 6.11:- DOA estimation of the three signals in the presence Mutual coupling.

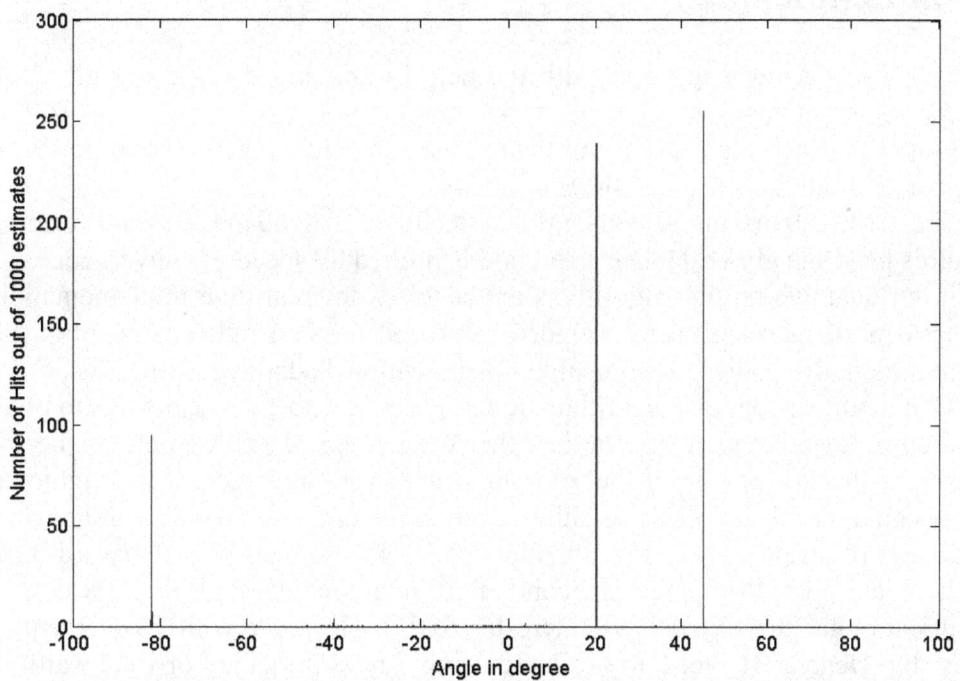

Figure 6.12:- DOA estimation of the three signals when Mutual coupling is compensated using open circuit voltage.

Note that, compensation of mutual coupling using open circuit voltages only reduces the effects of mutual coupling but not eliminates it.

Conclusions

The theoretical investigations, with the help of simulation results using MATLAB 6.5, presented in this book have clearly indicates that,

1) Simple null steering beam-former is not used in actual practice because of its large number of limitations listed in the section 3.8.1.

2) The LMS algorithm is conventional technique of adaptive beam-forming and requires pilot signal to calculate error, and is limited by speed of convergence.

3) Faster adaptive nulling algorithms are required for real time implementation. The Direct data domain least square algorithms based on the single snap shot of data are computationally efficient as compared to conventional adaptive algorithms.

Conventional adaptive algorithms require large number of snap-shots to obtain the solution of adaptive weights. On the other hand D^3LS algorithm obtains the solution of weights using single snap-shot of data. More over convergence of weight vector is always guaranteed for D^3LS algorithm, even in the presence of coherent interferers. A D^3LS algorithm such as, Forward method or Backward method or Forward-backward method has ability to suppress the coherent or non-coherent interferers, which may be 70 dB stronger than Signal Of Interest (SOI). However with 13 elements linear array, the Degree Of Freedom available using Forward method or Backward method is only 6. But using Forward-Backward method Degree of Freedom increases to 8 with slightly added computational complexity. In case of Forward method adaptive nulling algorithm using 13 elements linear antenna array , the Degree Of Freedom is 6 (5 interferers plus 1 SOI), so if number of interferers is greater than 5, then in the adapted beam-pattern only 5 nulls are formed corresponding to the 5 stronger interferers. But additional interferers (greater than 5) reduce the output SINR depending upon their strength towards poor value as described in section 4.4.4.

The only requirement of D^3LS algorithm is that, the direction of arrival of desired signal should known.

4) Conjugate Gradient Method is quite suitable for real time implementations of D^3LS algorithm. It is used to solve $[A][W]=[Y]$ equation in an iterative fashion. At the start more number of iterations may be required, but once the system adapts itself to the signal, it needs only one or two additional iterations per each new signal sample to obtain an acceptable solution.

5) Mutual coupling effect, adversely affect the nulling capability of adaptive nulling and DOA estimation algorithms. In the presence of mutual coupling, the nulls formed in the directions of interferers are shallow or sometimes nulls are misplaced. The open circuit voltage compensation technique can reduce the effect of mutual coupling effect but fails to eliminate it completely.

6) Direction Of Arrival estimation, based on the Matrix Pencil method is accurate, does not require multiple data snapshots, can handle coherent multipath and is faster as compared to popular techniques, such as Root-MUSIC. This gain in speed and reduced data requirements is due to Matrix Pencil not requiring an estimate of a covariance matrix. The only drawback is that, N+1 element array can estimate the DOA of N/2 signals only.

For example if an array consists of 13 elements, then Matrix Pencil DOA estimation algorithm can estimate DOA of 6 signals. When more than six signals impinge on the array, then DOA of six stronger signals are estimated. Moreover, with low Signal-to-Thermal noise ratio (18 dB), the estimate of DOA is superior when signals impinge on the array from close to broadside, as compared to end-fire case.

As we know in cellular mobile communications systems, major limiting factor is co-channel interference and fading. Fading is basically because of multipath nature of mobile channel. In order to combat co-channel interference and fading, adaptive antenna system is the best solution. Variety of adaptive nulling algorithms exists, but in mobile communications systems, we need such algorithms which are computationally efficient. Due to the time varying nature of mobile communication channel, weights in adaptive nulling algorithm need to be updated, based on the worst case fading rate.

Based on the conclusions arrived at each individual chapters of the book we find, Direct Data Domain Algorithms Based on the single snap-shot of data are most useful for real time implementation in mobile communications system. At present, it is feasible to implement adaptive antenna system at the base stations of cellular mobile systems rather than at the hand held mobile stations, because of present technology limitations. However, in future when technology will be advanced, it would be possible to implement adaptive antenna system in the handheld mobile stations too.

To make the D^3LS based adaptive nulling algorithms faster in operation, the Conjugate Gradient Method described in chapter 5 of thesis is quite useful.

For D^3LS based adaptive nulling algorithms, the main requirement is that, the direction of arrival of desired signal should be known. Faster estimation of direction of arrival can be easily carried out by using Matrix Pencil DOA estimation algorithm based on single snapshot of data.

www.ingramcontent.com/pod-product-compliance
Lightning Source LLC
Chambersburg PA
CBHW080702190526
45169CB00006B/2203